Collins

Cambridge Lower Secondary

Science

STAGE 7: WORKBOOK

Aidan Gill, Heidi Foxford,
Dorothy Warren

Collins

William Collins' dream of knowledge for all began with the publication of his first book in 1819.

A self-educated mill worker, he not only enriched millions of lives, but also founded a flourishing publishing house. Today, staying true to this spirit, Collins books are packed with inspiration, innovation and practical expertise. They place you at the centre of a world of possibility and give you exactly what you need to explore it.

Collins. Freedom to teach.

Published by Collins
An imprint of HarperCollins*Publishers*
The News Building
1 London Bridge Street
London
SE1 9GF

HarperCollins*Publishers*
Macken House, 39/40 Mayor Street Upper,
Dublin 1, D01 C9W8, Ireland

Browse the complete Collins catalogue at **www.collins.co.uk**

10 9 8 7

ISBN 978-0-00-836431-1

British Library Cataloguing-in-Publication Data

A catalogue record for this publication is available from the British Library.

End-of-chapter questions and sample answers have been written by the authors. These may not fully reflect the approach of Cambridge Assessment International Education.

This book is produced from independently certified FSC™ paper to ensure responsible forest management.

For more information visit:
www.harpercollins.co.uk/green

Authors: Aidan Gill, Heidi Foxford, Dorothy Warren
Development editors: Anna Clark, Lynette Woodward, Sarah Binns
Product manager: Joanna Ramsay
Content editor: Tina Pietron
Project manager: Amanda Harman
Copyeditors: Debbie Oliver, Naomi Mackay
Proofreader: Heather Addison
Illustrator: Jouve India Private Limited
Cover designer: Gordon Macgilp
Cover artwork: Maria Herbert-Liew
Internal designer: Jouve India Private Limited
Typesetter: Jouve India Private Limited
Production controller: Lyndsey Rogers
Printed and bound in India by
Replika Press Pvt. Ltd.

The publishers gratefully acknowledge the permission granted to reproduce the copyright material in this book. Every effort has been made to trace copyright holders and to obtain their permission for the use of copyright material. The publishers will gladly receive any information enabling them to rectify any error or omission at the first opportunity.

Acknowledgements

(t = top, c = centre, b = bottom, r = right, l = left)

p 36l Michael Potter11/Shutterstock, p 36r neelsky/Shutterstock, p 59 Hemanshu Gandhi/Shutterstock, p 135 Webspark/Shutterstock, p 137 Windows Original after Northcott, p 139tr, bl, br Designua/Shutterstock, p 162 Siberian Art/Shutterstock, p 163t vectortatu/Shutterstock, p 163bl THANAKRIT SANTIKUNAPORN/Shutterstock, p 163br Morphius Film/Shutterstock.

Contents

How to use this book

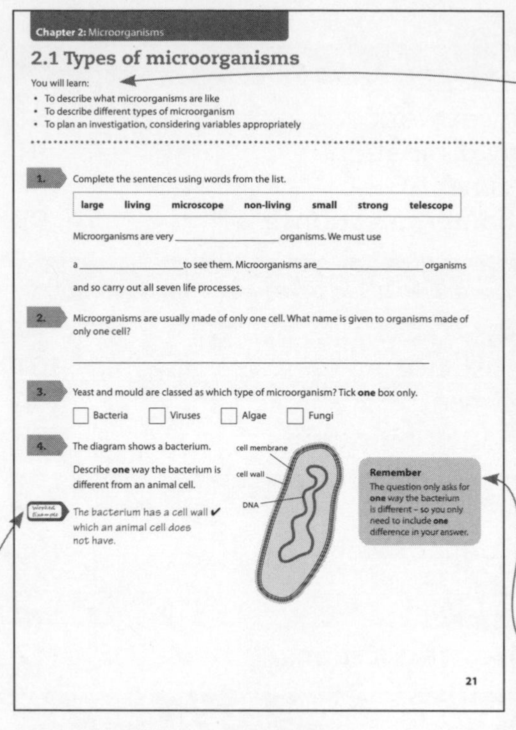

Chapter 2: Microorganisms

2.1 Types of microorganisms

You will learn:

- To describe what microorganisms are like
- To describe different types of microorganism
- To plan an investigation, considering variables appropriately

1. Complete the sentences using words from the list.

large	living	microscope	non-living	small	strong	telescope

Microorganisms are very ______________ organisms. We must use a ______________ to see them. Microorganisms are ______________ organisms and so carry out all seven life processes.

2. Microorganisms are usually made of only one cell. What name is given to organisms made of only one cell?

3. Yeast and mould are classed as which type of microorganism? Tick **one** box only.

☐ Bacteria ☐ Viruses ☐ Algae ☐ Fungi

4. The diagram shows a bacterium.

Describe **one** way the bacterium is different from an animal cell.

Worked Example: *The bacterium has a cell wall ✔ which an animal cell does not have.*

Remember
The question only asks for **one** way the bacterium is different – so you only need to include **one** difference in your answer.

21

These lists show what you will cover in the questions

Learn how to structure your answers with 'show me' questions where part of the answer is already completed for you

There are handy hints and tips in the 'remember' boxes

The worked examples will show you a sample answer

There are questions developing your practical skills throughout

Stretch yourself with these challenge questions

5. Explain why a fungus is **not** classed as a plant. Give an example of a fungus.

Show Me: *A fungus does not* ______________.

An example is ______________.

6. Aiko leaves some lychees in a bowl for two weeks.

After 10 days she observes patches of grey mould on the surface of the lychees. The mould grew from spores that landed on the lychees.

Suggest where these spores came from and how they got on the lychees.

7. Practical: Lana is looking at a book with a picture of some bacteria. The bacteria in the image are magnified ×10 000. What does this mean?

8. Challenge: Microorganisms are so small that we measure them in micrometres (μm).

One micrometre is one-thousandth of a millimetre.

a. Complete the table by converting the size of the yeast cell and virus particle into millimetres and the *Coccus* cell into micrometres.

Type of microorganism	Average diameter	
	Micrometres (μm)	Millimetres (mm)
Yeast cell	8	
Coccus cell		0.001
Virus particle	0.1	

b. Rank the three microorganisms in order of size from smallest to biggest.

22

2

End-of-chapter questions

1. a. Microorganisms are very small. Name the piece of apparatus needed to view microorganisms.

b. Which of the following statements about microorganisms is true?

Tick **one** box.

☐ Some microorganisms are decomposers
☐ All microorganisms are multicellular
☐ There are not many types of microorganisms
☐ Microorganisms cannot reproduce

2. Aafiya is investigating decomposers.

She places a slice of strawberry in a Petri dish as shown in the figure below:

slice of strawberry
petri dish
strawberry

Aafiya observes the surface of the strawberry every day for two weeks.

- On day 5 there is a small patch of mould on the surface of the strawberry.
- On day 14 the strawberry is completely covered in mould and has become smaller.

a. What type of organism is a mould? Tick **one** box only.

☐ Bacteria ☐ Fungus ☐ Virus ☐ Plant

b. What gas does the mould release during respiration?

c. Where does the mould on the strawberry come from?

31

At the end of the chapter, try the end-of-chapter questions! The questions will cover the topics within the chapter

At the end of the chapter, fill in the table to work out the areas you understand and the areas where you might need more practice. There is space for your teacher to comment too

Teacher's comments

30

Biology

1.1 Characteristics of living things

You will learn:

- To describe the characteristics used to identify living things
- To describe how scientists make hypotheses that are tested
- To choose appropriate ways to present results

1. Add ticks to the table to show whether the named things are living or non-living. The first row has been done for you.

	Living	Non-living
Rock		✓
Fungus		
Tree		
Star		

2. Draw **four** lines to match the life process to the correct description.

Life process	**Description**
Respiration	Getting bigger
Growth	Responding to changes in the environment
Sensitivity	Getting rid of wastes
Excretion	Providing energy

3. Name the life process you can see in the image.

____

Remember

Here's a tip to remember the life processes. Make a sentence using the first letter of each one. Or make a word or phrase using those letters. For example: MRS GREN.

4. Explain why movement is essential for survival of an antelope.

Worked Example

Movement allows an antelope to move away from danger, such as predators, ✔ so they do not get killed. ✔ It also allows them to find food ✔ so they have the right nutrition to stay alive. ✔

Remember

To answer this question you should give **two** reasons why movement is essential and then explain why for each one.

5. Bears have a very good sense of smell. Suggest **two** reasons why this could help them survive.

Show Me

A good sense of smell helps bears ______________________________

and ____ .

6. Raja says 'Respiration is the same as breathing'. Explain why this is **not** true.

Challenge

____

____

____

7. How would you present the following data? Choose from a bar chart or a line graph and explain your choices:

a Favourite foods in the class: ______________________________

____

b Changes in numbers of rabbits found living in an area over a year: ______________

____

8. Practical

Hitesh grows a radish plant. He measures the height of the plant every other day and records his results in this table.

Day	0	2	4	6	8	10	12
Height (cm)	0.0	0.2	0.7	1.6	2.2	2.7	3.0

a Complete the line graph below to show Hitesh's results.

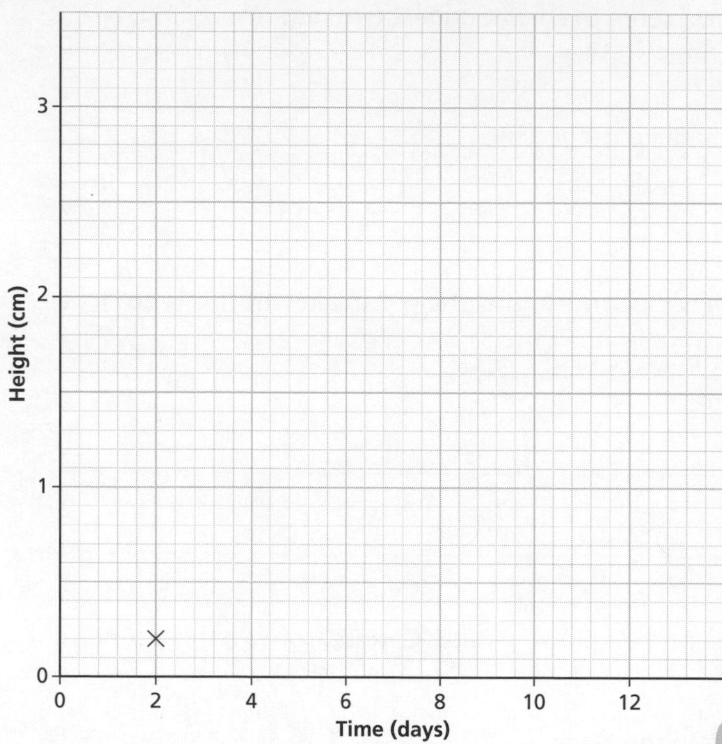

Remember
Use a sharp pencil to plot each point on the line graph with a small and clear 'x'.

b Is the radish a living organism? Use evidence from the graph to support your answer.

____

9. Practical

Fatma is investigating how calcium affects the strength of bones. Calcium is a mineral found in bones.

Fatma uses a chemical to remove the calcium from four bones.

She then tries to bend each bone to find out if it breaks easily.

a What question is Fatma trying to answer?

____

Fatma does some research and finds that calcium is needed for strong bones.

b Write a hypothesis for Fatma's investigation.

____

c Fatma finds that each of the four bones breaks very easily after the calcium has been removed. Does this evidence support your hypothesis?

____

1.2 Cells, tissues, organs and organ systems

You will learn:

- To understand how cells make up organisms
- To describe the relationship between cells, tissues, organs and organ systems
- To describe some organ systems and their functions

1. The human body contains cells. What are cells? Tick **one** answer only.

- ☐ Organs in our bodies that control what we do
- ☐ Tiny living units found in some parts of the body
- ☐ Tiny living units that the body is made from
- ☐ Very small round structures that are found only in the brain

2. Explain what a tissue is.

Show Me

A tissue contains one type of ______________________.

These work together to ______________________.

Remember
Tissues contain only **one** type of cell. Make sure you are clear about the difference between tissues, organs and organ systems.

3. Xylem cells are found in the stems of plants. Xylem cells transport water to the leaves. Explain why a group of xylem cells is considered to be a tissue and not an organ.

____

____

4. The diagram below shows major organs in the human body. Complete the labels A–D.

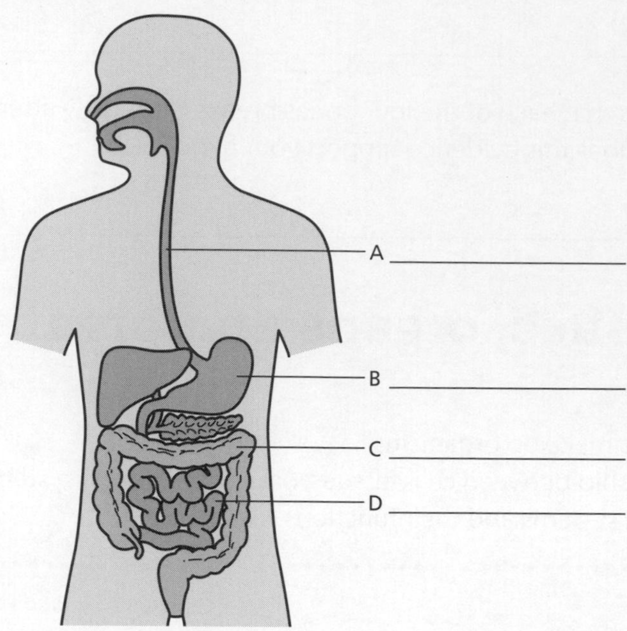

5. Draw **one** line to match each organ to the organ system it belongs to.

Organ	**Organ system**
Kidneys	Nervous system
Small intestine	Respiratory system
Brain	Excretory system
Lungs	Digestive system

6. Complete the table to show the function of the organs.

Name of organ	Function
Heart	
Bladder	
Diaphragm	

7. Name **two** organs in the circulatory system.

____

8. The respiratory system moves air into and out of our body.

Show Me

Describe the difference between the air that moves into the body and the air that moves out of the body.

The air that moves into our body contains more ____________ and ______ carbon dioxide.

The air that moves out of our body contains more ______________________ and ________ oxygen.

9. Explain why it is an advantage to have a pair of organs rather than a single organ.

Challenge

____

____

Remember

You may need to do some research to find out where each organ is and how they are connected.

10. Plants also have organs. A leaf is a plant organ. Describe the function of a leaf.

____

11. Explain why the roots of a plant are considered to be an organ.

____

12. Practical

a Draw an outline of the human body on a large piece of paper. Add the organs of the digestive system to your drawing. Include labels for mouth, gullet, stomach, pancreas, small intestine, large intestine and rectum.

b Find out what each organ does and add this information to your diagram.

c What do all of the digestive organs work together to do?

____

1.3 Comparing plant and animal cells

You will learn:

- To identify the main parts of cells and describe their functions
- To compare and contrast animal and plant cells
- To describe the difference between unicellular and multicellular organisms

1. The diagram shows a cell viewed under a light microscope.

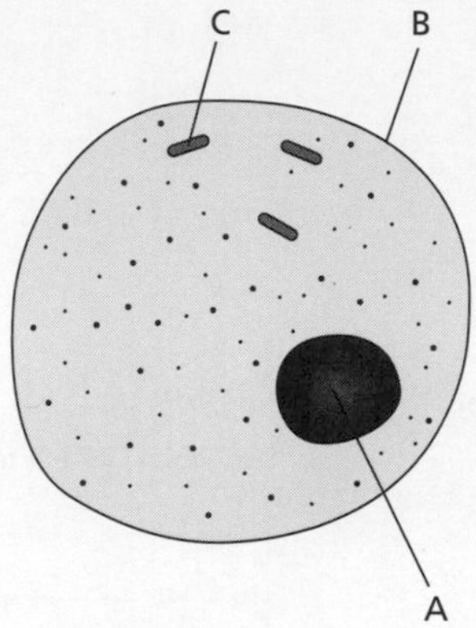

a Name the structures labelled A, B and C.

A ______________________

B ______________________

C ______________________

b Does the diagram show a plant or animal cell?
Explain your answer.

____

2. Add ticks to show which parts of a cell are found in plant cells, and which are found in animal cells. The first row has been done for you.

Part of cell	Animal cell	Plant cell
Cell membrane	✓	✓
Cell wall		
Cytoplasm		
Sap vacuole		
Mitochondria		

3. Explain why a blood cell can only be seen using a microscope.

Show Me

Each blood cell is ______________ in size.

A microscope ______________________________.

4. Describe the function of the following parts of a cell.

Cytoplasm ______________________________

Cell wall ______________________________

Chloroplasts ______________________________

5. Add ticks to show whether each named organism is unicellular or multicellular. The first row has been done for you.

Name of organism	Unicellular	Multicellular
Human		✓
Bacteria		
Plant		
Algae		

Remember

Organisms made of billions of cells are called multicellular organisms. Organisms made of only one cell are called unicellular organisms.

6. Draw a line to match each part of the microscope with its function.

Practical

Part of microscope	**Function**
Eyepiece lens	The part you turn to produce a clear image
Stage	The part you look through
Focusing wheel	The flat surface where you put the slide

7. Suggest why plant cells contain chloroplasts but animal cells do not.

Challenge

____

____

1.4 Specialised cells

You will learn:

- To identify some specialised cells
- To explain how some cells are adapted to certain functions

1. Complete the sentences using words from the list.

adapted	**cells**	**algae**	**faster**	**nuclei**	**unable**

Some ______________ are specialised. This means they have a shape or feature

that makes them ________________ to perform a particular function.

2. What do specialised cells work together to form? Tick **one** box.

☐ Larger cells ☐ Tissues

☐ Cytoplasm ☐ Organ systems

3. Draw **one** line to match each specialised cell to its function.

Name of specialised cell	**Function**
White blood cell	To carry oxygen around the body
Red blood cell	To carry electrical signals around the body
Neurone	To fight disease

> **Remember**
> The command word **describe** means you should state what something is like. **Explain** means you need to say why or how something happens, so use words such as 'because' or 'so'.

4. Describe and explain why a root hair cell is different to a cell found in a leaf.

Show Me

Challenge

A root hair cell has a ____.

These features maximise ____ into the root from the soil.

The leaf cell has ______________________________________ because these

are the parts of the cell that ______________________________________.

5. Challenge

Okan looks through a microscope at some cells from a potato tuber. Potato tubers grow under the ground.

Okan can see many dark circular structures inside each cell, but his teacher says these are **not** chloroplasts.

Explain why there are **no** chloroplasts.

____

____

____

6. Describe the function of a ciliated cell and explain how it is adapted to its function.

____

____

____

7. The diagrams below show different specialised cells.

A B C

For each specialised cell, explain one way in which the cell is adapted to carry out its job.

Cell A: ____

____

Cell B: ____

____

Cell C: ____

____

8. Aiko makes a model of a red blood cell using some modelling clay.

Practical

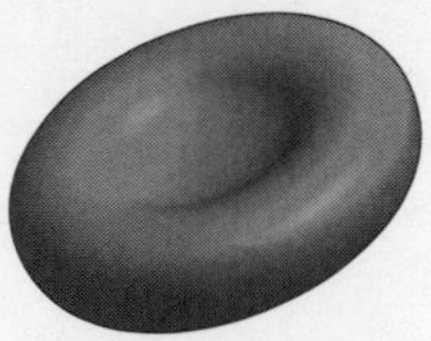

She gives each side of the red blood cell a bi-concave shape. Her model accurately represents the shape of red blood cells.

This bi-concave shape has more surface area and more cell membrane compared to just a flat disc.

The bi-concave shape gives red blood cells a large surface area.

a What advantage does the bi-concave shape give the red blood cell?

___

___

b Describe **one** other adaptation of the red blood cell.

___

___

Self-assessment

Tick the column which best describes what you know and what you are able to do.

What you should know:	I don't understand this yet	I need more practice	I understand this
All organisms carry out seven life processes so that they can survive			
The life processes are: movement, reproduction, sensitivity, growth, respiration, excretion, nutrition			
Scientists collect data to use as evidence for their hypotheses and to answer their questions			
All organisms are made of cells			
Cells form tissues, which form organs, which work together in organ systems in large organisms			
Main organs in the body include the skin, lungs, blood vessels, heart, diaphragm, liver, stomach, small intestine, large intestine, kidneys, bladder			
Many organs work together in organ systems, such as the circulatory system, the nervous system, the respiratory (breathing) system and the digestive system			
Plants have organs that include leaves, roots, stems and flowers			
Organisms range in size from unicellular organisms such as bacteria to multicellular complex organisms such as humans			
Animal cells have a nucleus, cytoplasm, mitochondria and a cell membrane			
Plant cells also have a sap vacuole and a cell wall, mitochondria and often have chloroplasts			
A light microscope is used to examine slides containing specimens			
Specialised cells have adaptations so that they can do certain jobs			

Specialised plant cells include palisade cells and root hair cells			
Specialised animal cells include neurones, blood cells and ciliated cells			
You should be able to:	**I can't do this yet**	**I need more practice**	**I can do this by myself**
Make careful observations including measurements			
Present results in the form of tables, bar charts and line graphs			
Spot patterns in data shown in tables, bar charts and line graphs			
Use a hypothesis to make a prediction			
Use data to either support or form evidence against a hypothesis			
Plan a simple investigation to test a hypothesis			
Choose appropriate apparatus and use it correctly (microscope)			
Interpret data from simple tables, bar charts and line graphs, including those presented in a graph, chart or spreadsheet			

If you have ticked 'I don't understand this yet' or 'I can't do this yet' or mostly 'I need more practice', have another look at the relevant pages in the Student's Book. Then make sure you have completed all the questions in this Workbook chapter and the review questions in the Student's Book. If you have already completed all the questions, ask your teacher for help and suggestions on how to progress.

Teacher's comments

....

End-of-chapter questions

1. This list shows things which some living organisms can do.

breathe	excrete	grow	make their own food	move	reproduce

a Choose **two** things a cow **can** do.

1. ______________________________

2. ______________________________

b Choose **one** thing a cow **cannot** do.

____

2. Paulo is looking at some cells under a microscope. In each cell he can see a nucleus, a cell wall and chloroplasts.

a Explain the function of each of these structures.

Nucleus: ______________________________

Cell wall: ______________________________

Chloroplasts: ______________________________

b Is Paulo looking at plant or animal cells? Explain your answer.

____

____

____

3. The diagram below shows a type of specialised cell.

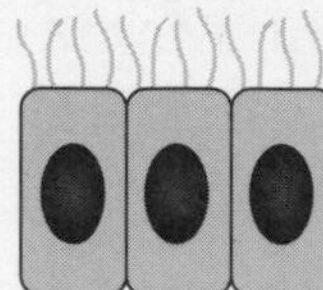

a Name the type of cell shown in the diagram.

____

b Explain **one** way this cell is adapted to its job.

____

4. A human body contains many types of tissues. Describe how a tissue is different from an organ.

____

____

____

2.1 Types of microorganisms

You will learn:

- To describe what microorganisms are like
- To describe different types of microorganism
- To plan an investigation, considering variables appropriately

1. Complete the sentences using words from the list.

large	**living**	**microscope**	**non-living**	**small**	**strong**	**telescope**

Microorganisms are very ____________________ organisms. We must use

a ____________________ to see them. Microorganisms are ____________________ organisms

and so carry out all seven life processes.

2. Microorganisms are usually made of only one cell. What name is given to organisms made of only one cell?

____

3. Yeast and mould are classed as which type of microorganism? Tick **one** box only.

☐ Bacteria ☐ Viruses ☐ Algae ☐ Fungi

4. The diagram shows a bacterium.

Describe **one** way the bacterium is different from an animal cell.

cell membrane
cell wall
DNA

Worked Example The bacterium has a cell wall ✔ which an animal cell does not have.

Remember
The question only asks for **one** way the bacterium is different – so you only need to include **one** difference in your answer.

5. Explain why a fungus is **not** classed as a plant. Give an example of a fungus.

Show Me

A fungus does not ______________________________.

An example is ______________________________.

6. Aiko leaves some lychees in a bowl for two weeks.

After 10 days she observes patches of grey mould on the surface of the lychees. The mould grew from spores that landed on the lychees.

Suggest where these spores came from and how they got on the lychees.

____

____

7. Lana is looking at a book with a picture of some bacteria. The bacteria in the image are magnified ×10 000. What does this mean?

Practical

____

8. Microorganisms are so small that we measure them in micrometres (μm).

Challenge

One micrometre is one-thousandth of a millimetre.

a. Complete the table by converting the size of the yeast cell and virus particle into millimetres and the *Coccus* cell into micrometres.

Type of microorganism	Average diameter	
	Micrometres (μm)	Millimetres (mm)
Yeast cell	8	
Coccus cell		0.001
Virus particle	0.1	

b. Rank the three microorganisms in order of size from smallest to biggest.

____

9. Zaafir wants to find out what conditions make mould on chapatti grow most quickly.

Practical

Zaafir thinks chapatti left in warm conditions will grow more mould than chapatti left in cool conditions.

Write a plan for Zaafir to find out which condition results in the most mould growth. Your answer should include:

a A list of the apparatus Zaafir will need

____

____

b Instructions on how Zaafir should use the apparatus

____

____

____

2.2 Microorgansims and decay

You will learn:

- To describe the ecological role of microorganisms as decomposers
- To construct and interpret food chains and food webs that include decomposers
- To describe the common stages of scientific investigations

1. Complete the sentences using words from the box.

air	eruption	food	generation	pasteurisation	water

Before the work of Louis Pasteur, scientists believed that microorganisms could arise from non-living matter. This was called spontaneous ____________________. Louis Pasteur discovered that there are microorganisms in the ____________________ that spoil food.

2. Practical

The diagram below shows the steps in one of Pasteur's investigations into what makes food go bad.

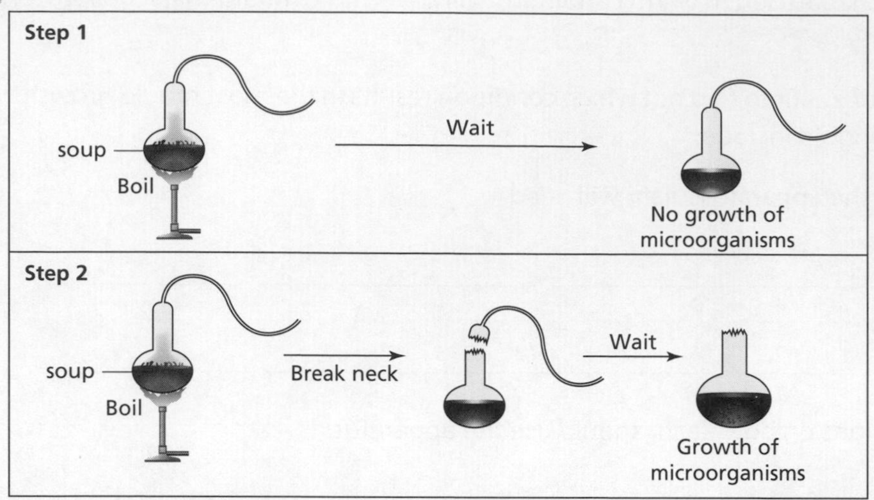

a What scientific question was Pasteur trying to answer? Tick **one** box only.

- [ ] Does soup contain microorganisms?
- [ ] Do microorganisms grow?
- [ ] Why does soup go bad?
- [ ] Does soup need to be boiled?

> **Remember**
> Evidence is data or observations we use to support or contradict a hypothesis.

b Explain why Pasteur boiled both soups at the start of his investigation.

____

c Why did Louis Pasteur use an S-shaped flask?

____

Show Me

d Pasteur concluded that microorganisms in the air caused the soup to go bad. What was the evidence for his conclusion?

The evidence was that only the soup in the flask

____.

Challenge

e Explain why Pasteur's findings were so important.

____

____

____

3. Why do fridges and freezers help to prevent food from going bad?

____

4. Mr Lee leaves some bread, dried fruit and pickled vegetables in his cupboard while he goes on holiday. The term 'pickled' means that the vegetables have been soaked and stored in vinegar.

When he gets back, the bread is mouldy. But the dried fruit and pickled vegetables are **not** mouldy. Explain why.

____

____

5. There are several stages in the scientific method used by scientists.

Practical

Complete the table by writing 1st, 2nd, 3rd or 4th to show the correct order of the stages shown.

Stage of scientific method	Order
Experiment	
Question to be answered	
Hypothesis	
Prediction	

6. Pasteurised milk eventually goes bad even if the container is not opened. Explain why.

Challenge

____

____

7. Draw lines to match each key term to its correct definition.

Key term	**Correct definition**
Prediction	An idea that can be tested to answer a scientific question
Hypothesis	Observations or data used to support or oppose an idea
Evidence	What you think will happen in an investigation

8. **Practical**

Bahula is investigating ways of stopping food going bad by using some fresh fruit and some salt.

She cuts two equally sized pieces of mango and rubs salt over one of the pieces. She then puts each piece into a different jar.

- **Jar 1:** Mango rubbed in salt
- **Jar 2:** Mango without salt

a What scientific question is Bahula trying to answer?

____

b Bahula writes a prediction: 'I predict the mango that is not rubbed in salt will go bad before the mango rubbed in salt.'

Write an explanation for Bahula's prediction.

____

____

c Bahula leaves the jars for two weeks. She then observes the two jars. What evidence might support her prediction?

____

____

9. Food goes bad when microorganisms cause it to decay. What is meant by the term 'decay'?

____

10. **Practical**

Petra wants to know what conditions cause bread to decompose more quickly. She has an idea that temperature affects how quickly mould can reproduce and decompose bread. She puts a slice of bread in each of two plastic bags.

- One bag is left in a refrigerator.
- One bag is left in a warm room.

Petra counts the number of mould colonies (spots) on each piece of bread after 10 days.

a What is Petra's hypothesis?

____

b Write a prediction for Petra's investigation.

___

c What evidence would Petra need to support the prediction you have written?

___

11. Decomposers are microorganisms that cause decay. Tick the boxes to indicate **two** decomposers.

☐ **viruses** ☐ **bacteria**

☐ **fungi** ☐ **spiders**

12. The figures below show four living things found in a grassland habitat.

snake grass hawk grasshopper

Not to scale

- grasshoppers eat grass
- snakes eat grasshoppers
- hawks eat snakes

Complete this food chain for the four living things shown.

Grass ⟶ ___ ⟶ ___ ⟶ ___

13. Look at the organisms in question 12. Complete the table below by giving the names of **one** predator, **one** prey, **one** producer and **one** primary consumer.

Predator	Prey	Producer	Primary consumer

14. Which type of organism uses the Sun's energy to make its own food?

☐ **consumer** ☐ **producer** ☐ **herbivore** ☐ **predator**

15. Draw lines to match each word with its meaning.

Word	**Meaning**
Herbivore	animal that eats both plants and animals
Carnivore	animal that eats plants
Omnivore	animal that eats other animals

16. A food chain shows 'what eats what'.

What do the arrows in a food chain represent?

____

17. Juan says: 'All the energy in a food chain comes from the Sun.'

Worked Example

Explain why this statement is true.

Energy from the Sun is stored in plants during photosynthesis (1). This energy is passed along the food chain when one organism consumes another (1).

> **Remember**
> A food chain describes the order in which organisms depend on each other for food.

18. Explain the role of decomposers in a food chain.

____

____

19. Not all the energy in producers reaches the primary consumers, and not all the energy in the primary consumers reaches the secondary consumers. Explain why.

Challenge

____

____

Self-assessment

What you should know:	I don't understand this yet	I need more practice	I understand this
Microorganisms are living things that you need a microscope to see			
Most microorganisms are single celled			
The three main types of microorganism are viruses, bacteria and some fungi			
Louis Pasteur did experiments to gather evidence to support his idea that microorganisms make food go bad			
Some microorganisms (such as some bacteria and fungi) are decomposers			
Decomposers cause the decay of dead organisms and animal waste, clearing these away and releasing mineral salts back into the soil			
Food chains and food webs show how energy passes from one organism to another in a habitat			
Decomposers form a part of all food chains and food webs			

You should be able to:	I can't do this yet	I need more practice	I can do this by myself
Find, use and evaluate information from different sources			
Choose equipment to use and use it correctly			
Make careful observations and measurements			
Identify and design scientific questions, testable hypotheses and predictions			
Identify evidence used to make conclusions			
Recall the different stages commonly used in scientific investigations			

If you have ticked 'I don't understand this yet' or 'I can't do this yet' or mostly 'I need more practice', have another look at the relevant pages in the Student's Book. Then make sure you have completed all the questions in this Workbook chapter and the review questions in the Student's Book. If you have already completed all the questions, ask your teacher for help and suggestions on how to progress.

Teacher's comments

End-of-chapter questions

1. **a** Microorganisms are very small. Name the piece of apparatus needed to view microorganisms.

____

b Which of the following statements about microorganisms is true?

Tick **one** box.

- [ ] Some microorganisms are decomposers
- [ ] All microorganisms are multicellular
- [ ] There are not many types of microorganisms
- [ ] Microorganisms cannot reproduce

2. Aafiya is investigating decomposers.

She places a slice of strawberry in a Petri dish as shown in the figure below:

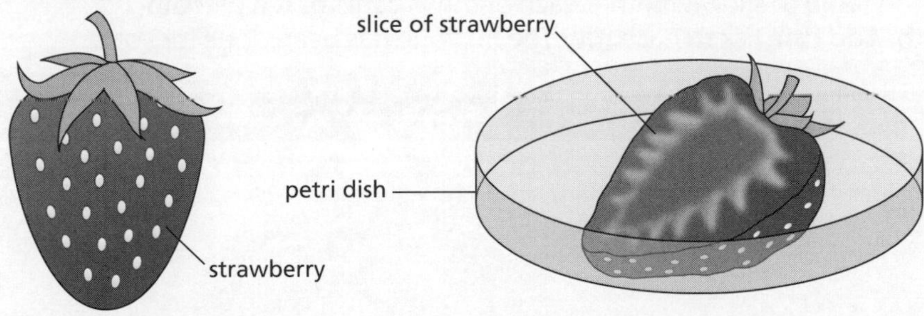

Aafiya observes the surface of the strawberry every day for two weeks.

- On day 5 there is a small patch of mould on the surface of the strawberry.
- On day 14 the strawberry is completely covered in mould and has become smaller.

a What type of organism is a mould? Tick **one** box only.

- [ ] Bacteria
- [ ] Fungus
- [ ] Virus
- [ ] Plant

b What gas does the mould release during respiration?

____

c Where does the mould on the strawberry come from?

____

d Explain why the strawberry becomes smaller.

e Aafiya does some research and is surprised to learn that mould is not a type of plant. Explain **one** main difference between mould cells and plant cells.

3. The diagram shows a food chain for organisms that live in an area of farmland.

alfalfa plant → aphid → beetle → bat

Not to scale

a Complete the table to show whether each of the organisms is a predator, prey or both. Add **one** tick to each row. The first one has been done for you.

Animal	Predator	Prey	Both
aphid		✓	
beetle			
bat			

b Name a herbivore shown in the food chain.

c In this food chain, the alfalfa plant is the producer. What is meant by 'producer'?

d Many microorganisms live in the soil of the farmland. Microorganisms are important for healthy growth of the alfalfa plants. Explain why.

3.1 Classifying organisms

You will learn:

- To use sources of information to classify organisms based on their characteristics
- To describe what a species is
- To explain why there is no virus kingdom

1. Complete the sentences by choosing words from this list.

characteristics	height	invertebrates	kingdoms	species

Living organisms can be classified according to their ______________________.

Organisms are classified into large groups called ______________________ such as 'animal', 'plant' and 'fungi'.

2. Draw lines to match each organism with the kingdom to which it belongs.

Organism	**Kingdom**
Crab	fungi
Moss	animal
Mushroom	plant

3. Humans are part of the animal kingdom. Explain why humans are **not** classified as plants.

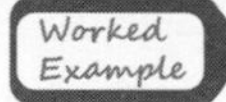

Humans are not classified as plants because human cells do not have cell walls (1). Also, human cells never contain chlorophyll and so they cannot make their own food like plant cells (1).

> **Remember**
> The plant kingdom contains multicellular organisms that contain chlorophyll and make their own food by photosynthesis.

4. Who invented the system of classification we use to classify organisms? Tick **one** box.

- [ ] **Charles Darwin**
- [ ] **Francis Crick**
- [ ] **Carl Linnaeus**
- [ ] **Robert Bakewell**

5. Animals can be classified as vertebrates or invertebrates. Describe the main difference between vertebrates and invertebrates.

Show Me

A vertebrate is ______________________________.

An invertebrate is ______________________________.

6. The images show some different organisms.

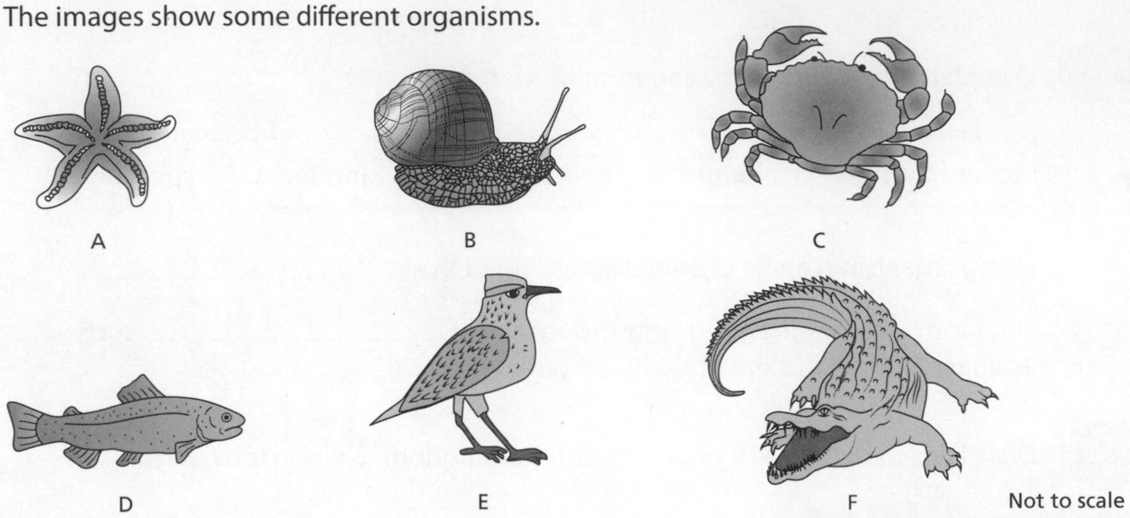

A

B

C

D

E

F

Not to scale

Remember

You should know the characteristics of the vertebrate groups: mammals, fish, amphibians, reptiles and birds.

a Give the letter of the organism that lays eggs with leathery shells.

____

b List the letters of the images that are invertebrates.

____

c Which letter shows an arthropod?

____

d Give **two** characteristics of birds.

1. ______________________________

2. ______________________________

7. The pictures below show two different types of pony.

a Describe a **similarity** in the characteristics of the two ponies.

____

b Describe a **difference** in the characteristics of the two ponies.

____

c What term is used to describe the differences in the characteristics between the two ponies?

____

8. Describe what is meant by the term 'species'.

Worked Example

A group of organisms that can reproduce with one another ✔ and have offspring that can also reproduce. ✔

9. The African elephant and Sumatran tiger are both mammals but are two different species.

Suggest two differences between the two mammals to show they are different species.

____

____

10. A zonkey is a hybrid created by interbreeding a zebra with a donkey.

a What is a hybrid?

____

b Suggest why zonkeys are rarely found in the wild.

____

11. Challenge

Viruses cannot replicate without invading a 'host' cell. Viruses are **not** put into any of the kingdoms. Explain why.

____

____

12. Challenge

Aristotle was a Greek scientist who developed the first classification system. He divided all the known organisms into two groups – the plants and the animals.

Aristotle then divided each animal group into three smaller groups – land, water and air. Birds, bats and flying insects were grouped together as 'air animals'.

We now know that Aristotle's classification system does **not** work. It groups organisms that are completely different to one another in the same group.

Explain why Aristotle's classification of 'air animals' does **not** work.

____

____

3.2 Biological keys

You will learn:

- To use and construct biological keys for classification

1. Why do scientists use classification keys?

____

2. A student finds some organisms living in the sea.

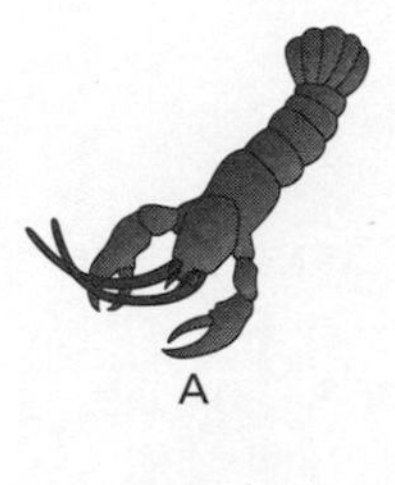

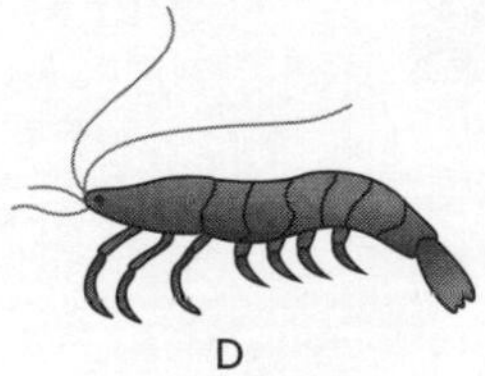

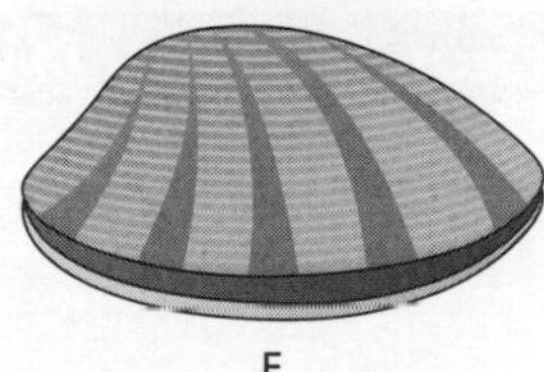

F

Not to scale

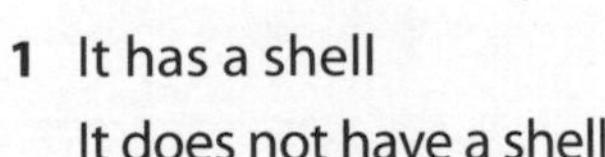

1	It has a shell	Go to 2
	It does not have a shell	Go to 3
2	Shell has one piece	Whelk
	Shell has two pieces	Clam
3	It is divided into segments	Go to 4
	It is not divided into segments	Go to 5
4	It has claws	Lobster
	It does not have claws	Shrimp
5	It has five arms	Starfish
	It has more than five arms	Octopus

Use the dichotomous key to answer these questions.

a Which letter shows a whelk? ______________________

b Which letter shows a clam? ______________________

c What is organism A? ______________________

d What is organism D? ______________________

3. Fatima collects leaves from four different types of tree. She is going to make an identification key so that each type of tree can be identified using its leaves.

Show Me

Describe what Fatima needs to do to make an identification key.

> **Remember**
> Identification keys have questions that help you identify living organisms.

Look carefully at the ________________________.

Write ________________________ based on the characteristics.

4. Jalal finds some organisms in the school grounds. The diagrams below show the organisms he finds.

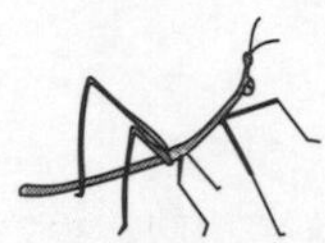

stick insect

jumping spider

meal moth

hoopoe bird

Not to scale

Jalal makes a table to record the characteristics of the organisms he finds.

Name of organism	Number of legs	Number of wings
Stick insect	6	0
Jumping spider		
Meal moth		
Hoopoe bird		

a Use the diagrams to complete the table. The first one has been done for you.

Challenge **b** Jalal uses the information in his table to construct a classification key. Complete the key by writing the missing question.

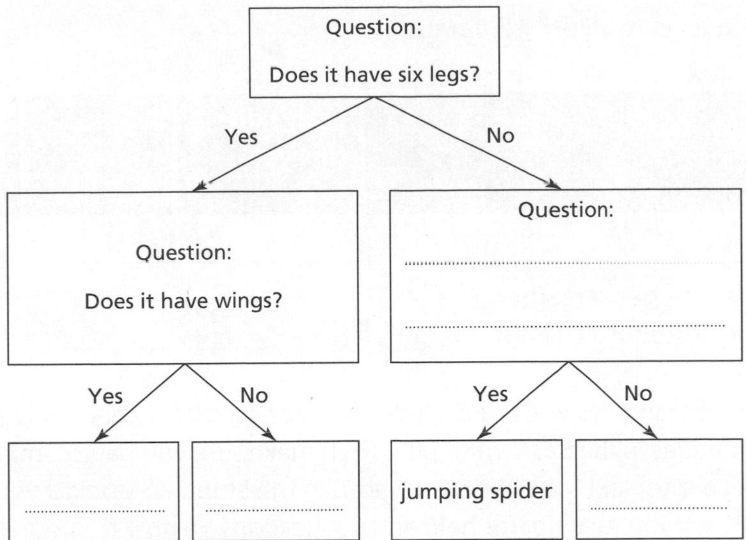

Challenge **c** Write the name of each organism in the table in the correct box on the key above. One has been done for you.

Self-assessment

Tick the column which best describes what you know and what you are able to do.

What you should know:	I don't understand this yet	I need more practice	I understand this
Scientists classify organisms in groups by looking at their characteristics			
The largest classification groups are kingdoms			
The animal kingdom contains five groups of vertebrates – mammals, reptiles, fish, amphibians and birds			
The animal kingdom contains vertebrates and invertebrates, both of which contain smaller groups			
The plant kingdom contains groups such as flowering plants and conifers			
The smallest group in a kingdom is the species, which is a group of very similar organisms that can reproduce to produce fertile offspring			

Viruses cannot replicate without being inside a living cell and so are not considered to be living			
A dichotomous key can be used to identify different plants and animals			
You should be able to:	**I can't do this yet**	**I need more practice**	**I can do this by myself**
Make careful observations			
Use similarities and differences between groups of items to use and create dichotomous keys			

If you have ticked 'I don't understand this yet' or 'I can't do this yet' or mostly 'I need more practice', have another look at the relevant pages in the Student's Book. Then make sure you have completed all the questions in this Workbook chapter and the review questions in the Student's Book. If you have already completed all the questions, ask your teacher for help and suggestions on how to progress.

Teacher's comments

End-of-chapter questions

1. The pictures below show four different organisms.

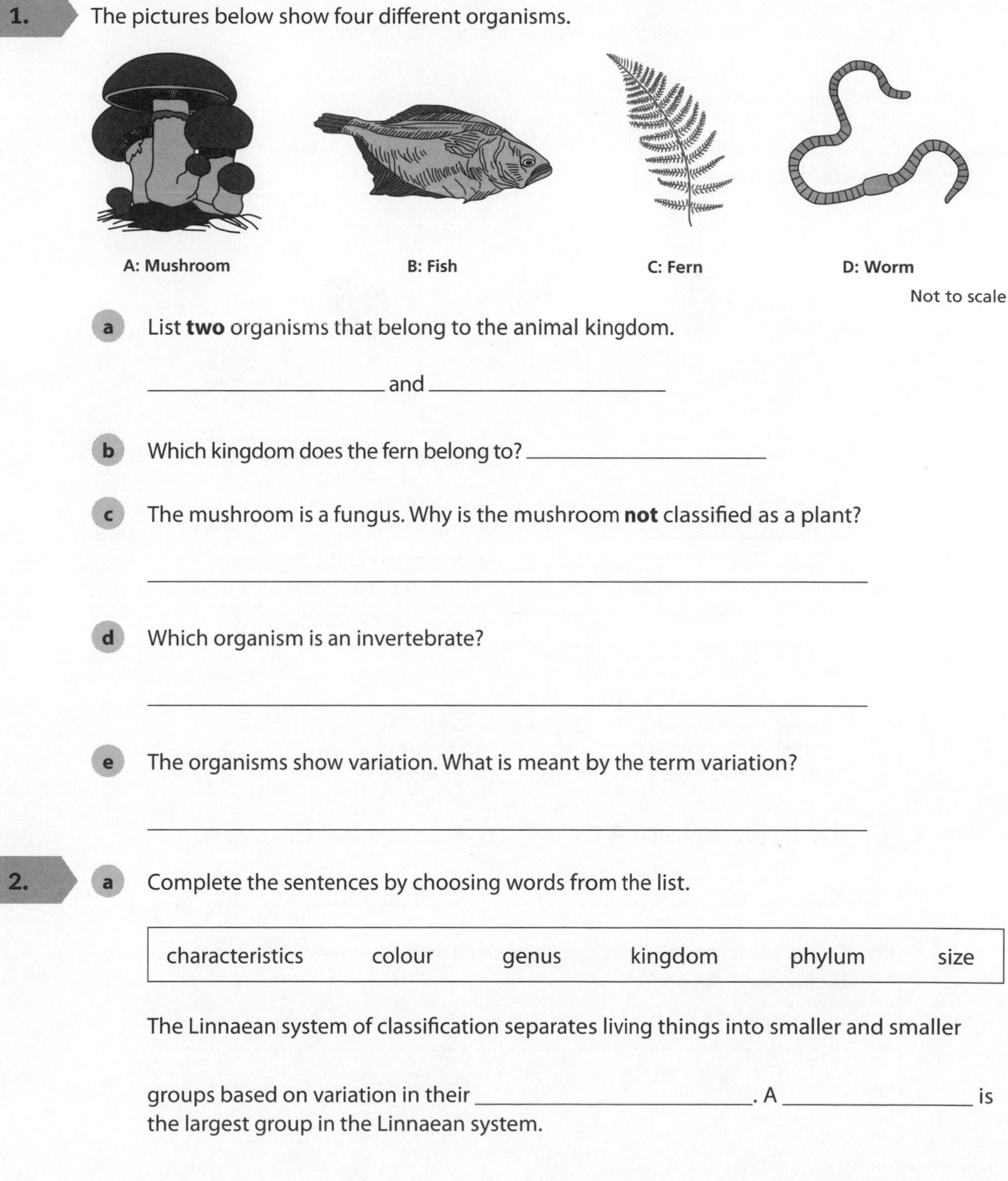

a List **two** organisms that belong to the animal kingdom.

____________________ and ____________________

b Which kingdom does the fern belong to? ____________________

c The mushroom is a fungus. Why is the mushroom **not** classified as a plant?

____

d Which organism is an invertebrate?

____

e The organisms show variation. What is meant by the term variation?

____

2. **a** Complete the sentences by choosing words from the list.

characteristics	colour	genus	kingdom	phylum	size

The Linnaean system of classification separates living things into smaller and smaller

groups based on variation in their ____________________. A ________________ is the largest group in the Linnaean system.

b Explain why there is not a kingdom for viruses.

____

c Give one way in which viruses are similar to living organisms.

____

3. Surika finds some insects.

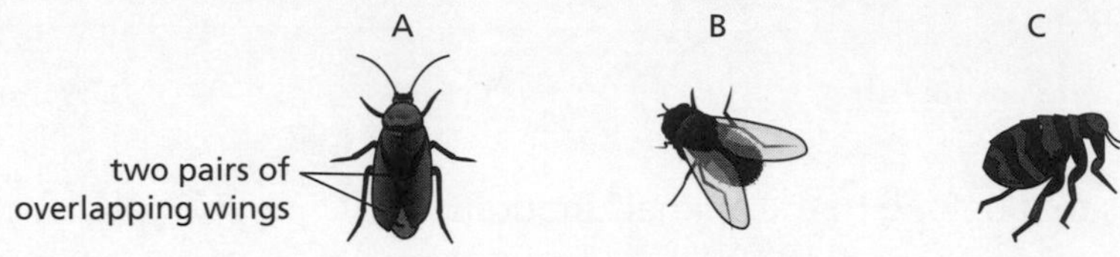

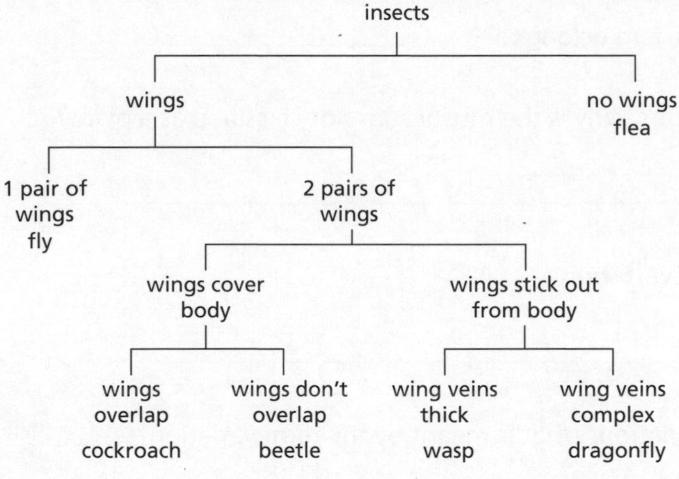

a Use the key to identify **A**, **B** and **C**.

A: ____________________

B: ____________________

C: ____________________

b Surika's teacher says insects are classified as invertebrates. What is meant by 'invertebrate'?

____

Chemistry

4.1 Physical and chemical properties

You will learn:

- To understand the chemical and physical properties of a substance
- To plan an investigation, considering variables appropriately
- To choose appropriate apparatus and use it correctly
- To describe trends and patterns in results, including identifying any anomalous results
- To make conclusions by interpreting results

1. Complete the sentences.

Physical properties are those that can be observed and ______________ such as mass, conductivity and boiling points.

Chemical properties are those that can be seen when a substance takes part in a chemical

______________ such as flame colour or temperature change.

2. The box contains some words/phrases used to describe some of the properties of different substances.

pH	malleable	melting point	reactivity	boiling point
	hardness	flammability	heat of combustion	

Sort the words/phrases into physical and chemical properties.

The first one has been done for you.

Physical property	Chemical property
Boiling point	Heat of combustion

3. Pure ice melts at 0 °C. At what temperature does water freeze?

Tick **one** box

–10 °C	0 °C	10 °C	100 °C
☐	☐	☐	☐

4. Look at the diagram.

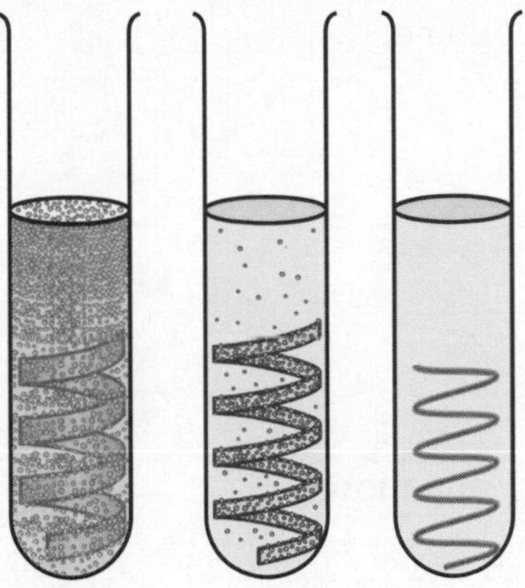

Circle the chemical property that is being investigated.

flame colour	reactivity	temperature change	pH

5. Look at the table of melting point and boiling point data.

Substance	Melting point (°C)	Boiling point (°C)
Water	0.0	100.0
Salt	801.0	1413.0
Carbon dioxide	–56.6	–78.5
Olive oil	–6.0	300.0

a What state of matter is water at 330 °C? ______________________________

b What state of matter is carbon dioxide at 20 °C? ______________________________

c What state of matter is salt at 100 °C? ______________________________

d What state of matter is olive oil at –10 °C? ______________________________

6. Practical

Savita and Padma are planning an investigation to see how the temperature of ice changes as it is heated to its boiling point.

a What variable should they control?

____

Remember

When using negative numbers, a 'higher number' means a number with a lower value. For example, –10 has a lower value than –5.

b What will they record in their results table?

Write the column headings and units in the table outline.

c Describe what the girls will observe during the investigation.

Show Me

Solid pieces of ice will get smaller until. ______________________

As the ______________________ small bubbles will be seen to get bigger on the surface of the pan. ______________________ will rise from the pan.

7. Practical

Ershad and Vijay carried out an investigation to see how the temperature of liquid wax changed as it cooled down.

They plotted their results in this graph:

Remember
Describe means to write about what you **see** without giving any reasons.

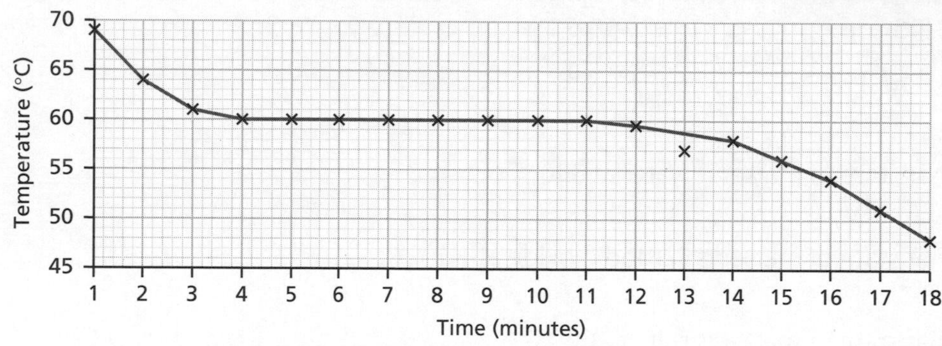

a Describe what happens to the temperature of the wax during the experiment.

____

____

b Draw a ring around the data point on the graph that does **not** fit the pattern of the other measurements.

c What temperature did their wax freeze at? ______________________

8. This question is about investigating solubility.

Practical

Angelique and Anastasia make three different solutions.

The girls added a spoonful of each solid substance to half a cup of water and stirred.

When all the solid had dissolved they add another spoonful of solid and continued to stir.

This step was repeated until no more solid would dissolve.

Here are the results

Salt – 10 spoons

Flour – 1 spoon

Coffee granules – 3 spoons

a Write down the independent variable.

____

b Write down the dependent variable.

____

c State a variable that Angelique and Anastasia should control.

____

d Complete the results table. You will need to include a heading in each column.

e Use the results to compare the solubility of each substance.

____

____

____

____

f Describe **two** pieces of equipment the girls could use to make up their solutions more accurately.

___

___

g Is solubility a physical or chemical property? Give a reason for your answer.

___

___

4.2 Acidity and indicators

You will learn:

- To understand acidity and alkalinity and how it is measured
- To recall and use hazard symbols
- To evaluate experiments and investigations, and suggest improvements, explaining any proposed changes

1. Add ticks to the table to show whether a substance is acidic, alkaline, or neutral. The first row has been done for you.

	Acidic	Alkaline	Neutral
Lemon juice	✓		
Toothpaste			
Distilled water			
Vinegar			

2. State the meaning of each of the four hazard symbols shown below.

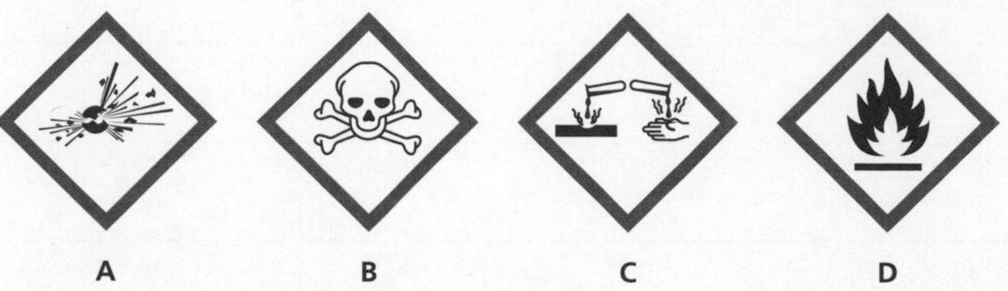

3. Use words from the box to label the diagram of the pH scale.

neutral	acidic	alkaline	weak	strong	pH

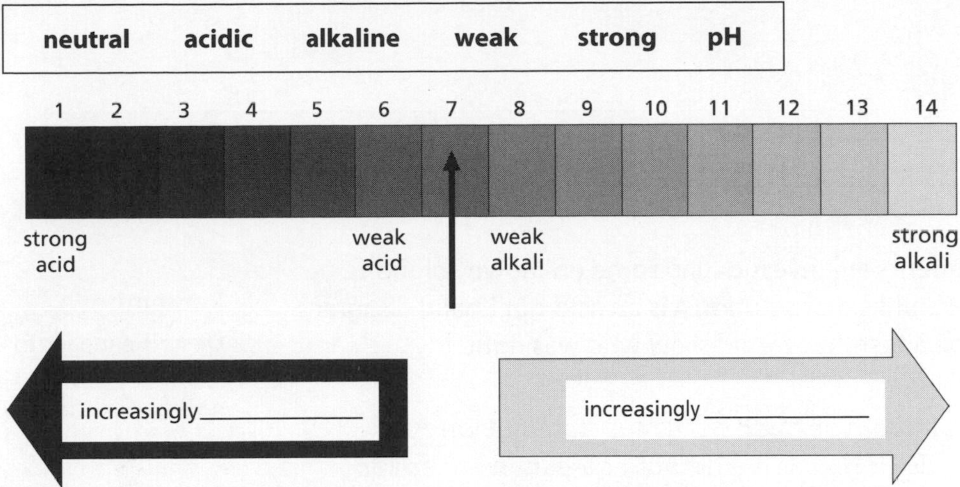

4. **a** Why do some chemicals have hazard symbols?

____

b Caustic soda or sodium hydroxide is used to make soaps and other cleaning products. Look at the hazard symbol on the label.

When handling this product what precautions should be taken?

____

____

c Suggest a reason why the label has a hazard symbol rather than lots of writing.

____

5. Universal Indicator is sometimes described as a 'full range indicator'.

Explain why.

____

____

6. Two students are investigating some unknown solutions. Rhadish thinks that solution A is an acid but Khalid disagrees. Describe a test that would show who was right.

Remember
Describe means to 'give an account of'. In this case you must include the expected result.

Worked Example

Add an indicator such as litmus to solution A. ✓
If the solution goes to its acid colour, in this case red, then it is an acid. ✓

7. Heather plants grow best in acidic soils but other plants such as clematis prefer alkaline soils. Describe how you would decide which soil sample, A or B, was suitable for growing heathers.

Remember
A four-mark question needs four different points to be made in the answer.

Show Me

I would carry out a ______________ on the soil sample.

First, I would mix the soil with water. Then I would ______________ and ______________. Finally I would observe the colour change and ______________.

8. Explain why Universal Indicator paper goes dark green when toothpaste is added but orange when lemon juice is added.

____

____

9. Explain why Universal Indicator is more useful than other indicators such as litmus.

Challenge

____

____

____

10. **Practical**

This question is about making and using natural indicators. Asef has been learning about indicators at school and wants to try and make his own at home. He gathers together some different foods that he is going to make the indicators from.

This is his method:

- chop up the food (if needed)
- mix it with hot water
- filter it

- add some of the filtered indicator solution to a known acidic solution and also a known alkaline solution
- record the results.

a Describe how Asef could improve his method to ensure a fair comparison between indicators.

____

____

____

Asef was so pleased with the results that he took his indicators to school to do some more tests.

Here are his results.

The results of Asef's experiments are shown in the table:

Indicator	Colour in acids	Colour in neutral	Colour in alkalis
Beetroot	Red	Pink	Purple
Curry powder	Pale yellow	Yellow	Red
Turnip skin	Red	Blue	Green

b When Asef added oven cleaner to the curry powder indicator, it went red.

What can he conclude about the pH of oven cleaner?

____

c Milk has a pH of about 6.5.

Look at the indicators shown in part (a). State which indicator you would use to test the pH of milk to make sure it had not gone off. Explain your answer.

____

____

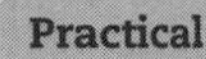

You might like to have a go at making your own indicator at home. Many coloured flowers also make good indicators! Simply follow the method at the start of the question. For the filtering step, you could use a kitchen sieve. Test your indicator with materials such as soap or fruit juices.

4.3 The particle model

You will learn:

- To use the particle model to describe different states of matter
- To describe a vacuum
- To describe the strengths and limitations of a model

1. Add ticks to the table to show whether a substance is a solid, liquid or gas at 20 °C (room temperature). The first row has been done for you.

	Solid	Liquid	Gas
Tomato sauce		✓	
Bubbles in a glass of fizzy drink			
Oil			
Sand			

2. Draw **three** lines to match the particle diagram to the correct state of matter.

Particle diagram	**State of matter**
	liquid
	gas
	solid

Describe how you could use building bricks to model the arrangement of particles in a solid.

____

____

____

3. Look at the data in the table.

Substance	Properties
A	It is hard, has a fixed shaped and is very strong
B	It flows, can be compressed and fills the container
C	It flows, cannot be compressed and takes on the shape of the container

a Which substance is a liquid? ______________________

b Give a reason for your answer. ______________________

____

____

4. Use the particle model to explain why a gas can be compressed but a liquid cannot.

Worked Example

The particles in a gas are far apart, so when a pressure is applied they are brought closer together. ✓ The particles in a liquid are already touching each other, so when a pressure is applied they cannot get any closer. ✓

Remember

Explain why means you have to link what happens (or what does not happen) to a reason.

5. Use the particle model to explain why liquids can be poured but solids cannot.

Show Me

The particles in a liquid can move past each other because

____.

The particles in a solid cannot move because ______________________.

Write down a strength and a limitation of the particle model.

Strength ____

Limitation ____

6. Challenge

Rice is a solid but it shows some properties similar to those shown by liquids. Give an example of such a property and explain how this property can occur.

Remember

To answer this question you must apply your knowledge of the particle model.

____

____

____

7. The diagram shows some of the particles found in the Earth's atmosphere.

A B C D

a Which diagram represents atoms of argon gas? ______________

b Give two reasons for your answer.

____

____

c Which diagram represents particles of liquid water? ______________

d Give **two** reasons for your answer.

____

____

e Outer space does not have an atmosphere.

Explain why outer space is sometimes described as a vacuum.

____

____

4.4 Elements and the Periodic Table

You will learn:

- To understand how atoms make up all matter
- To know how elements are presented in the Periodic Table
- To represent the differences between elements, compounds and mixtures using the particle model
- To represent scientific ideas using symbols and formulae

1. Which statement is **not** true?

Tick **one** box.

- [ ] All elements are listed in the Periodic Table
- [ ] Each element is made from one type of atom
- [ ] Each element contains many different types of atoms
- [ ] All matter is made up of atoms

2. Draw a line to match the chemical symbol to its name

C Ca K P

Potassium Carbon Phosphorus Calcium

3. Crack the code.

Write down the element symbols to find the hidden word.

a chlorine oxygen carbon potassium

____

b carbon hydrogen iodine sodium

____

4. Explain why chemists find the Periodic Table so useful.

Show Me

Use words or phrases from this list to complete the answer.

grouped together	**don't have to remember them**			
randomly placed	**evaluate**	**predict**	**consider**	**in order**

The Periodic Table lists the names and symbols of all known

elements, so chemists ______________________ Elements

with similar properties are ______________________.

This allows chemists to ______________________ how different elements will react.

Remember

'**Explain**' means that you should say what is useful and **'why'** it is useful. Where there are three marks you should make three points.

5. Look at the diagrams A, B and C.

A B C

State which diagram represents an element. Give a reason for your answer.

____

____

6. Explain why chemists use chemical symbols.

Challenge

____

____

7. Use the clues provided to complete the crossword puzzle.

You may need to carry out some research to answer the questions.

Activity focused

Across

2 The element's chemical symbol is Ar

4 The element's chemical symbol is Mg

5 About 21% of the Earth's atmosphere

6 About 78% of the Earth's atmosphere

7 Often used to make rings and jewellery

9 This gas helps clean swimming pools

10 The element's chemical symbol is Be

12 This gas can be used to fill party balloons

Down

1 This element is commonly found as soot

3 You can find all the elements listed here

8 The simplest particle in the particle model

11 This element is used to make bridges

4.5 Elements, compounds and mixtures

You will learn:

- To describe how elements, compounds and mixtures are different
- To represent elements, compounds and mixtures using the particle model
- To represent scientific ideas using symbols and formulae

1. Draw a line to match the key word(s) to its diagram.

Element	Mixture of elements	Compound	A mixture of compounds

2. Identify each of the substances below as an element, compound or mixture.

Oxygen gas ____________________ Carbon dioxide gas ____________________

Water ____________________ Orange juice ____________________

Carbon ____________________ Air ____________________

3. Name the elements present in the compound sodium chloride.

____

4. Josh and Adam were discussing sea water.

Show Me

Adam said that sea water was a compound but Josh disagrees. He thinks that it is a mixture.

Explain why Josh is correct.

Pure water is a ____________________ made of two hydrogen atoms joined to one oxygen atom. Sea water is a mixture because it contains ____________________ as well as water particles.

5. Many musical instruments such as trumpets or cornets are made from brass.

Brass is an alloy of copper and zinc.

a State the meaning of the word **alloy**.

____

b Select the correct word in each sentence below to compare the differences in chemical and physical properties for brass as an alloy and copper and zinc as elements.

Brass is more/less reactive with water than copper or zinc.

Brass is harder/softer than copper or zinc.

Brass is weaker/stronger than copper or zinc.

Brass has a lower/higher melting point than copper or zinc.

c The particle model can be used to represent brass.

Look at the diagrams and identify which one best represents brass.

Tick **one** box

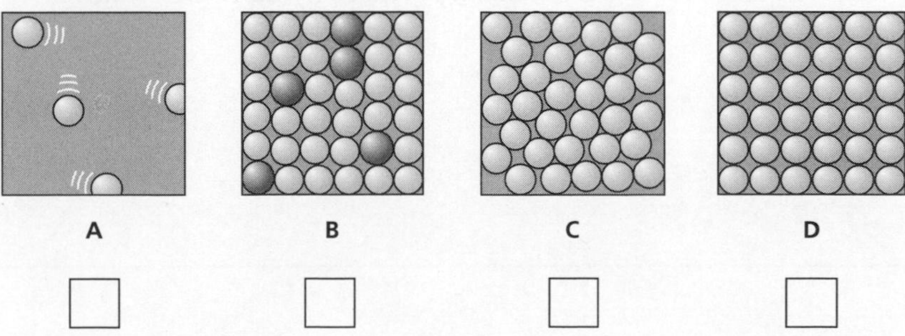

d Write down a strength and a limitation of the particle model you have chosen.

Strength ____

Limitation ____

6. Challenge

Copper sulfate and diamond both form crystals.

The formula of copper sulfate is $CuSO_4.5H_2O$.

The formula of diamond is C.

Why is copper sulfate a compound and diamond an element?

____

____

____

____

7. Ershad and Pramod were heating copper metal using a Bunsen burner. At the end of the experiment, they notice that the copper had gone black. Ershad thinks that the metal is covered in soot but Pramod disagrees.

Explain why Ershad is wrong.

____

____

____

8. Challenge Practical

A dish contains a black and yellow powder, which you predict is a mixture of iron and sulfur.

When a magnet is moved over the dish, black particles fly up to the magnet leaving a yellow powder behind.

State if the results support your prediction.

Explain the results of this test.

____

____

____

____

9. Investigating a mixture

Activity focused

Two girls were investigating mixtures. They decided to test the following hypothesis:

As the temperature of the water increases, the amount of substance dissolved in the mixture also increases.

Here is their method:

- Measure out 25 cm^3 water into an insulated beaker
- Measure the temperature
- Weigh out 150 g of sugar
- Add a spatula of sugar to the water in the beaker
- Stir until all the sugar has dissolved
- Add another spatula of sugar and stir
- Repeat adding sugar and stirring until no more sugar will dissolve
- Weigh the remaining sugar and work out how much has been added
- Repeat at four more temperatures

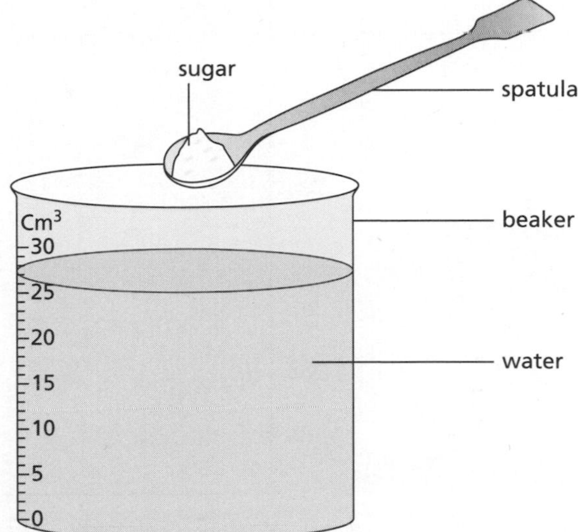

Temperature (°C)	Mass of sugar added (g)
20	50
40	60
60	73
80	88
100	120

a Use graph paper to plot a graph with mass of sugar added on the *y*-axis.

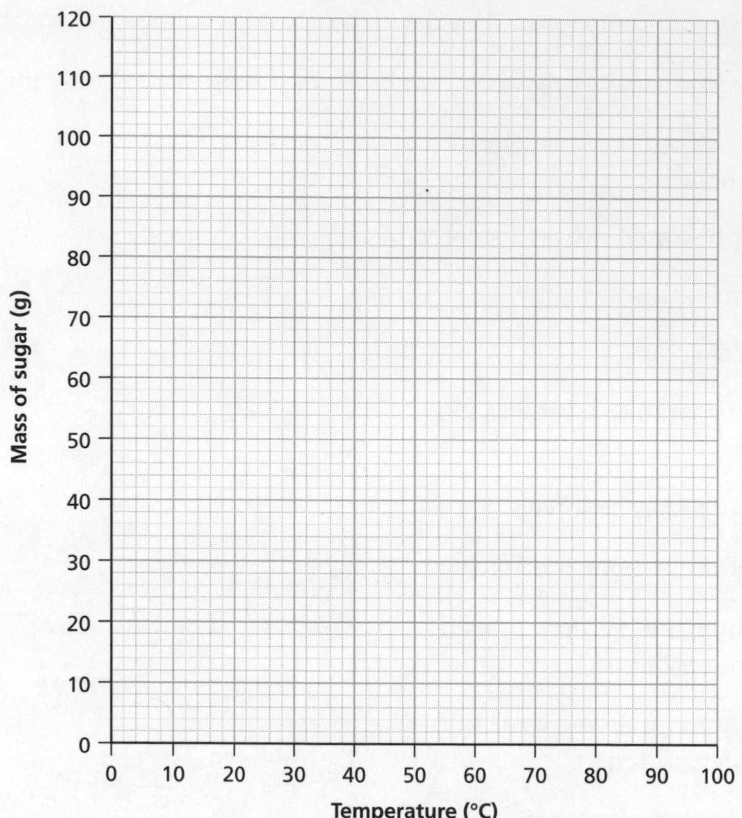

b Explain if their hypothesis is correct. Your answer must be supported by evidence.

____

____

____

10. A mixture contains 100 g of water and 10 g of salt.

State the mass of salt left if all the water evaporates. Explain your answer.

____

____

____

4.6 Properties of metals, non-metals and alloys

You will learn:

- To know how elements are grouped as metals and non-metals
- To describe the differences in physical properties between metals and non-metals
- To understand how alloys are different from their constituent substances
- To use the particle model to explain the differences in hardness between pure metals and their alloys
- To make conclusions by interpreting results

1. Complete the sentences using words from the list below.

more	**less**	**solids**	**liquids**	**gases**	**weaker**	**stronger**	**one**
			two	**three**	**states**		

At 20 °C most metals are ____________ but

non-metals will be found in all three ____________.

Metals are usually ____________ dense than non-metals.

Metals are usually ____________ than non-metals which can be easy to break.

2. The diagram shows the outline of the Periodic Table.

a Shade in the metals in grey.

b Give two physical properties of metals.

c Draw a green box around any non-metal.

3. This table lists the melting points of some metals and non-metals.

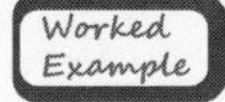

Substance	Metal/alloy	Melting point (°C)
Copper	Metal	1085
Tin	Metal	232
Lead	Metal	327
Solder	Alloy (tin and lead)	185
Bronze	Alloy (copper and tin)	850–1000

Use the data given in the table to compare the melting points of metals and their alloys.

Solder is an alloy of tin and lead. Its melting point of 185 °C is lower than that of both of the metals it is made from. ✓

> **Remember**
> **Compare** means to describe the similarities and differences of at least two things.

Bronze is an alloy of copper and tin. It has a broad melting range from 850 to 1000 °C, which is lower than the melting point of copper but greater than that of tin. ✓

Therefore, the data shows that alloys have different melting points to the elements they are made from. ✓

4. The table below shows some data for Substance A.

Substance	Melting point	Conducts electricity?	Hard or soft?
A	High	Yes	Soft

A student suggests that substance A is diamond.

Use the data in the table to decide whether the student is correct. Explain your answer.

> **Remember**
> When asked to 'use the data' from the table, you must refer to it in your answer.

____

____

____

5. Jack thinks that metals such as iron and copper are harder than non-metals such as graphite.

Practical

Lily says he should do a scratch test to see if he is right.

In a scratch test objects made of a harder material will scratch objects made of a softer material.

Explain how Jack can test his prediction.

____

____

6. Metals are often described as being malleable but many non-metals are not.

Challenge

Explain the meaning of 'malleable' and why non-metals do not have this property. Your answer should include examples.

7. **Investigating the strength of a metal wire**

Practical

The diagram shows a suspension bridge.

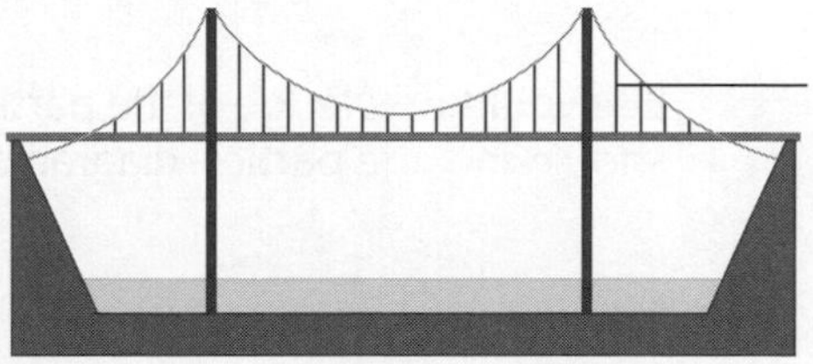

steel suspension cables

Nasser thinks that the cables on a suspension bridge are made from steel and not iron because steel is stronger than iron.

To test out his theory, Nasser plans an investigation using the equipment shown on the right.

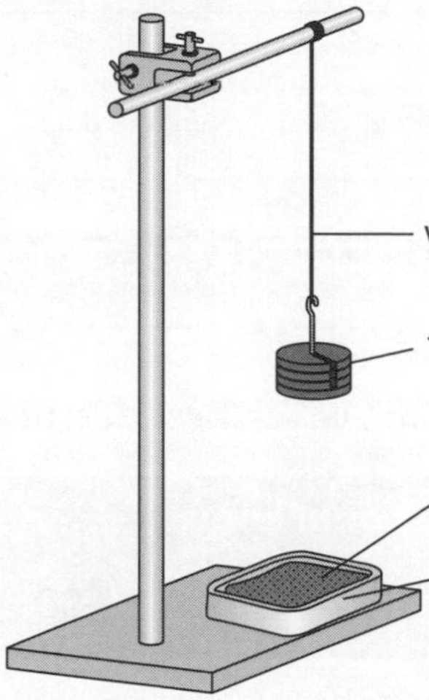

He added 100 g masses until the wire broke and then recorded the maximum mass the wire held.

wire being tested

100 g masses

sand

container

Here are his results:

Metal	Total mass added (g)	
	1st attempt	2nd attempt
Iron	800	900
Steel	1600	1400

a Explain why Nasser put a container full of sand under the masses.

b State **two** variables he needed to control during the investigation.

c Explain why Nasser tested each metal twice.

d Calculate the mean strength of each wire:

Iron: ____________________

Steel: ____________________

How much stronger is steel than iron?

____

____

e Use your knowledge of the particle model to explain why steel is harder than iron. You should include particle diagrams in your answer.

____

____

____

Self-assessment

Tick the column which best describes what you know and what you are able to do.

What you should know:	I don't understand this yet	I need more practice	I understand this
All substances have chemical and physical properties			
Physical properties are those that can be observed and measured such as mass, conductivity and melting and boiling points			
Chemical properties are those that can be seen when a substance takes part in a chemical change such as reactivity and flammability			
The acidity or alkalinity of a substance is a chemical property			
Acids and alkalis can be found in many everyday substances			
The pH scale goes from 0 to 14 and shows how acidic or alkaline a solution is			

The closer to 0 the pH is, the more acidic that solution is			
The closer to 14 the pH is, the more alkaline the solution is			
Neutral solutions have a pH of 7			
An indicator changes colour in a substance if it is acidic, alkaline or neutral			
Substances are made up of tiny particles			
The three states of matter are solids, liquids and gases			
Solids, liquids and gases have different properties because their particles are arranged differently.			
A vacuum is space entirely without matter			
Elements only contain one type of atom			
Each element has a symbol			
Elements are arranged in the Periodic Table in an order			
Compounds are formed when atoms of two or more elements combine in a chemical change			
Each compound has a formula, which shows you how many of each atom are in the compound			
The properties of a compound are different to the elements it contains			
A mixture is made up of at least two different elements or compounds			
Alloys are mixtures of metals			
Materials can be grouped as metals or non-metals			
Metals are usually hard, strong, malleable, and good conductors of heat and electricity			
Non-metals are usually dull, not shiny, and often have low melting and boiling points			

You should be able to:	I can't do this yet	I need more practice	I can do this by myself
Present observations and measurements in a suitable way			
Describe trends and patterns in results, including identifying anomalous results			

Interpret results and explain what they show			
Understand hazard symbols and consider them when doing practical work			
Plan investigations and make predictions based on scientific knowledge and understanding			
Decide what equipment is needed to carry out an investigation			
Group substances as solids, liquids and gases based on descriptions			
Use symbols to represent different elements in the Periodic Table			
Sort and group elements through testing and observations			
Sort, group and classify materials using observations and secondary information			
Use symbols and formulae to represent elements and compounds			
Make conclusions by interpreting results			
Evaluate experiments and suggest improvements			

If you have ticked 'I don't understand this yet' or 'I can't do this yet' or mostly 'I need more practice', have another look at the relevant pages in the Student's Book. Then make sure you have completed all the questions in this Workbook chapter and the review questions in the Student's Book. If you have already completed all the questions, ask your teacher for help and suggestions on how to progress.

Teacher's comments

End-of-chapter questions

1. This question is about elements, mixtures and compounds.

a Draw **one** line to match each element with its correct chemical symbol.

Helium	B
Hydrogen	Be
Beryllium	H
Boron	He

b Complete the sentences.

Choose from the words in the list.

element	**mixture**	**compound**

When iron and sulfur are mixed together and heated

they form a new____________________.

Bronze is an alloy. An alloy is a ____________________ of two or more different metals or a mixture of a metal with a non-metal element.

2. Look at the image. It shows that the colour of the hydrangea flower depends on the pH of the soil.

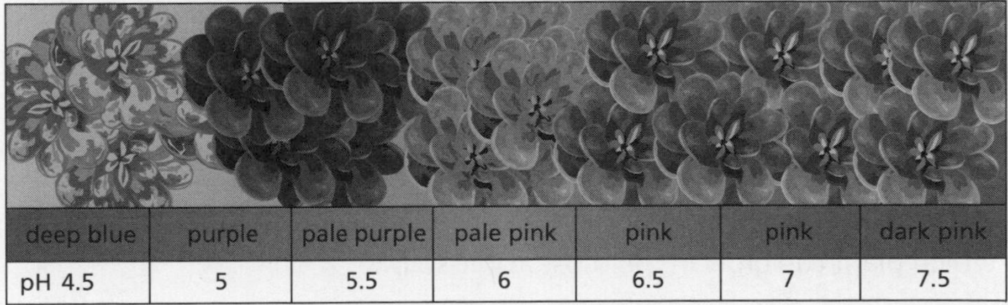

deep blue	purple	pale purple	pale pink	pink	pink	dark pink
pH 4.5	5	5.5	6	6.5	7	7.5

a The hydrangea flowers in the park are deep blue. State the pH of the soil.

____

b State the colour of the flowers of a hydrangea planted in neutral soil.

____

c Explain why hydrangea flowers change colour in different soils.

____

3. This question is about changes of state.

a Complete the sentence:

When a substance changes from a solid to a liquid, it ______________

b In the box, draw how the particles would appear in oxygen gas.

Draw at least **five** particles. The first particle has been drawn for you.

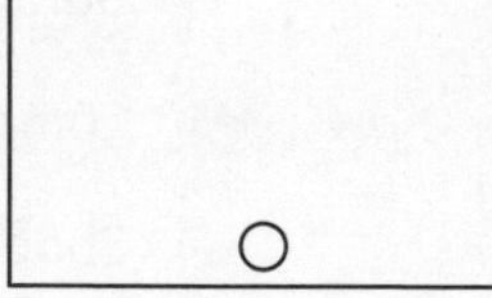

c Look at the diagram on the right. It shows the melting point and boiling point of bromine.

(i) What is the state of bromine at 65 °C? ______________

(ii) What is the temperature range of liquid bromine?

____

4. The table shows the soil pH conditions that some plants prefer to grow in.

Plant	pH of soil
Banana tree	5.5–6.5
Blueberry	4.5–5.4
Mango	6.2–7.5
Strawberry	5.0–7.0

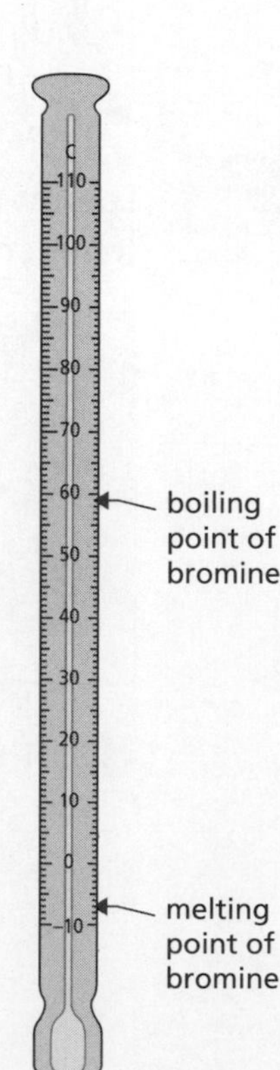

a State which plant can grow in the most acidic soil.

____

b State which plant grows well over the largest range of pH values.

____

c The pH of the soil can be tested by adding a small amount of the soil to water and adding a few drops of a suitable indicator.

Name the indicator that could be used.

____

5. Copper is a malleable metal which is sometimes used to make cooking pans.

a Describe what is meant by the term 'malleable' and explain why metals are malleable.

____

b State **one** other physical property and **one** chemical property of copper that make it a suitable material for cooking pans.

____

c Suggest **one** other use of copper and state the physical property that makes it suitable for this use. ____

6. This question is about gases.

a When air is pumped into a tyre, the tyre inflates (gets bigger) as the air fills it.

Which statement about gases is correct?

Tick **two** boxes.

- ☐ The particles in a gas expand
- ☐ The particles in a gas have fixed positions
- ☐ Gases have no fixed shape
- ☐ The particles in a gas only vibrate
- ☐ Gases have no fixed volume

b Describe what will be observed if all the air is removed from the tyre.

____

c Explain your answer to (b).

____

____

5.1 Making compounds

You will learn:

- To use observations to identify chemical reactions
- To identify oxygen, carbon dioxide and hydrogen gases using tests
- To describe chemical reactions using the particle model
- To represent scientific ideas using symbols and formulae

1. Draw a line to match the key word(s) to its diagram.

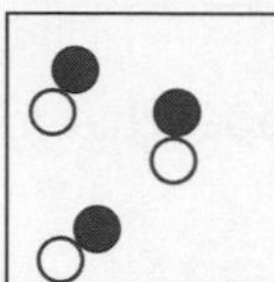
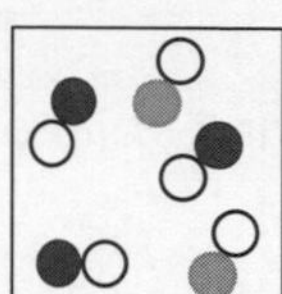
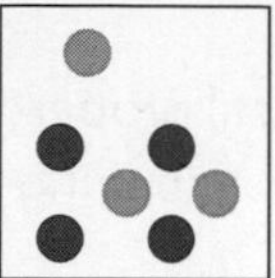
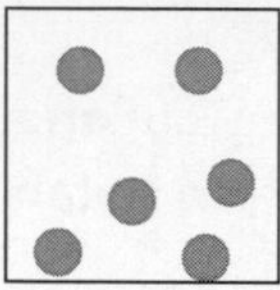

Element	Mixture of elements	Compound	A mixture of compounds

2. Complete the sentences by choosing the correct words from the list below.

compound	**nitrogen**	**air**	**oxygen**	**element**	**oxide**	**metal**
		nitrate	**mixture**			

When copper burns with a greeny-blue flame, it reacts with ____________________

to form a new ____________________ called copper ____________________.

3. Draw a line to match the test with the gas.

Turns limewater milky	Hydrogen
Burns with a squeaky 'pop'	Oxygen
Re-lights a glowing spill	Carbon dioxide

4. Look at the word equation and the particle model of the reaction between sulfur and oxygen.

sulfur + oxygen → sulfur dioxide

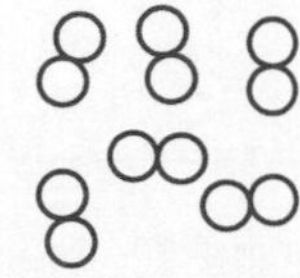
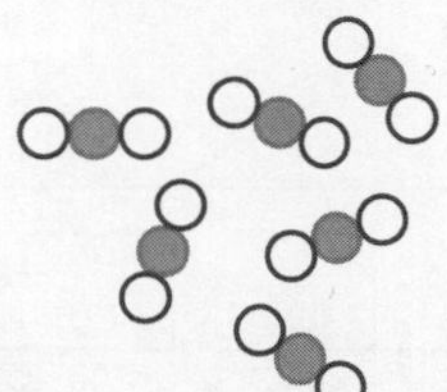

a Complete the sentences.

Choose from these words

solid	liquid	atom	elements	oxygen	gas	mixture

Sulfur and oxygen are both ______________.

An ______________ of sulfur reacts with ______________ gas to form sulfur dioxide ______________.

b Use the particle model to describe what happens during the reaction.

Show Me

Oxygen particles collide with ______________ particles. The particles ______________ and a new compound called sulfur dioxide is formed.

c Write down a strength and limitation of using the particle model to represent a chemical reaction.

Strength ____

Limitation ____

5. When magnesium burns in air the mass of the solid increases.

The word equation for the reaction is

magnesium + oxygen → magnesium oxide

a Use the particle model to explain why the mass of the solid increases.

____

____

____

____

b 6 g of magnesium reacts with 4 g of oxygen.

Calculate the mass of magnesium oxide formed.

____

6. When Shilpa adds some baking powder to vinegar she observes some fizzing.

Shilpa thinks that oxygen gas has been formed but her friend Vimla says that it is carbon dioxide gas.

Describe two simple tests the girls can carry out to see who is right.

____

____

____

____

7. Challenge

Two boys were heating metals using Bunsen burners. At the end of the experiment, they notice that the copper had gone black, but the iron nail looked just the same.

Explain their observations.

____

____

____

8. Practical Activity focused

Candle investigation

Mr Yung's class are investigating candles. He sets up the following equipment and asks the students to make a prediction.

Mingxia thinks the candles will burn for the same time because the beakers are the same size.

Elizaveta thinks the shortest candle will go out first because the carbon dioxide produced will sink to the bottom of the beaker and put the flame out.

a Explain why Elizaveta thinks that the carbon dioxide will sink.

____

b Explain why Mr Yung chose beakers that were the same size.

____

c Each group timed how long it took for the lighted candles to go out.

Here are the results from Mingxia's group.

Candle length in cm	Time candle burned in seconds			
	Experiment 1	Experiment 2	Experiment 3	Mean
10.9	26.0	22.0	18.0	22.0
6.3	22.0	23.0	17.0	
2.4	19.0	25.0	21.0	

Complete the table by working out the mean results. The first one has been done for you.

Here is the working: (26+22+18)/3 = 66/3 = 22

d Write a conclusion by comparing the means of the results.

____

e Comment on the quality of the data collected in the different experiments.

____

____

f Suggest **two** ways you could improve the experiment.

____

____

Try collecting some more data at home and see if it supports either Mingxia or Elizaveta's prediction.

5.2 Forming precipitates

You will learn:

- To explain how chemical reactions produce precipitates
- To describe chemical reactions using the particle model
- To represent scientific ideas using symbols and formulae
- To use scientific knowledge and understanding to make predictions

1. Draw a line to match the key word(s) to its definition.

Soluble	A substance that dissolves
Insoluble	A mixture of a dissolved substance and a liquid
Solution	A substance that does not dissolve

2. Identify the insoluble substances.

Tick **two** boxes.

☐ Silver chloride ☐ Calcium carbonate

☐ Sodium nitrate ☐ Hydrochloric acid

☐ Potassium hydroxide

3. Complete the sentences by choosing the correct words from the list below.

insoluble	**solution**	**product**	**precipitate**
compound	**soluble**	**mixture**	

Remember
When writing a word equation, you will find the answer in the question.

When ______________ substances react together to form an

insoluble ______________, a ______________ is formed.

4. When sodium hydroxide is added to copper sulfate, you can see a blue solid (copper hydroxide) forming in colourless sodium sulfate solution.

Worked Example

Write a word equation for the reaction.

sodium hydroxide + copper sulfate ✔ → copper hydroxide + sodium sulfate ✔

5. A white precipitate is formed when sodium chloride reacts with silver nitrate.

Show Me

a Write the word equation for the reaction.

potassium chloride + silver nitrate → ______________ + ______________

b Name the insoluble product ______________

c Use the particle diagrams to describe what happens during the reaction.

____

____

____

____

d Write **two** limitations of the particle diagram shown in part (a).

Limitation 1 ______________

Limitation 2 ______________

e Suggest how you could improve the particle diagrams.

____

____

Remember
When explaining something you must give a reason why.

6. Soluble mercury compounds sometimes get into the water supply.

Challenge

Explain why at some water treatment plants aluminium sulfate is added to the water flowing through the plant.

____

____

____

____

7. Challenge

Use your scientific knowledge and understanding to predict what will be observed when sodium iodide is added to silver nitrate.

You may wish to include particle diagrams in your answer.

____

____

____

8. Practical Activity focused

Investigating chemical reactions

Uri was carrying out some chemical experiments. He has noted down his observations in a table.

Test	Method	Observations	Conclusion
1	Add 6 drops of copper sulfate to sodium hydroxide	A blue precipitate is formed	
2	Add 6 drops of sulfuric acid to sodium hydroxide	No change	
3	Mixes some powered calcium carbonate with water	A cloudy white liquid	
4	A few drops of silver nitrate is added to sodium chloride	A white solid is seen at the bottom of the test tube	

a Complete the table.

Write a conclusion for each experiment based on the observations.

b Write a word equation for:

Test (1)

__________ + __________ → __________ + __________

Test (4)

__________ + __________ → __________ + __________

c Describe a further experiment Uri could carry out to provide further evidence to support a conclusion for test (3).

____

____

____

5.3 Neutralisation reactions

You will learn:

- To understand what neutralisation means
- To describe the change in pH during neutralisation
- To select and use the correct apparatus
- To safely carry out practical work

1. Use words from the box to complete the sentence.

acid	**alkaline**	**neutral**	**chemical**	**combustion**	**neutralisation**

A ______________ reaction occurs when acid is added to ______________ substances.

2. Name the salt produced when hydrochloric acid reacts with sodium hydroxide.

Tick **one** box.

- [ ] Sodium sulfate
- [ ] Sodium chloride
- [ ] Calcium chloride
- [ ] Calcium nitrate

3. In areas of hard water, limescale is deposited inside kettles when the water is boiled. Vinegar or lemon juice can be used to remove the limescale.

What does this tell us about the limescale?

Tick **one** box.

- [ ] It is neutral
- [ ] It is acidic
- [ ] It is basic (alkaline)
- [ ] It is very reactive

Remember
Explain means you have to say **what** happens and **why** it happens by giving a reason.

4. Show Me

John has just been stung by a bee. He wants to put some vinegar (mild acid) on it to ease the pain but his Mum says no, that will make it worse. She says he should use baking powder (mild alkali).

Explain why Mum is correct.

Bee stings are ____________________. To ease the pain the sting needs to be ____________________ by reacting it with a mild ____________________.

5. Challenge

If alkali A has a pH value of 13 and alkali B has a pH value of 9, explain which is most likely to be used in indigestion tablets as a remedy for excess stomach acid.

____

____

____

6. Practical

This question is about neutralisation.

A student is planning an investigation to find out the pH of a range of household substances to determine whether each is acidic, alkaline or neutral.

The five substances to be tested are:

Lemon juice

Pure water

Shampoo

Milk

a Name a suitable indicator that could be used in this investigation and describe how it shows the pH of each substance to be determined.

____

____

b Suggest a safety precaution the student must take during this investigation.

____

c Give two ways in which the student can make sure this is a fair test.

____

____

d The table below shows the pH values for each substance.

Substance	pH	Acidic, alkaline or neutral
Lemon juice	2	
Pure water	7	
Shampoo	8	
Bleach	14	

e Use the data provided in the table to determine whether each substance is acidic, alkaline or neutral. Explain your answer.

____

____

____

____

Self-assessment

Tick the column which best describes what you know and what you are able to do.

What you should know:	I don't understand this yet	I need more practice	I understand this
Word equations and particle diagrams are used to model chemical reactions			
How to identify whether a chemical reaction has happened by the loss of reactants or the formation of products			
Compounds can be made during chemical reactions			
Products have different properties to the reactants that formed them			
Evidence that a chemical reaction has happened includes the production of a gas or a precipitate, or a change of colour			

There are specific tests to identify hydrogen, oxygen and carbon dioxide			
Soluble substances can dissolve in water			
Insoluble substances cannot dissolve in water			
A precipitate is an insoluble product that forms in the reaction between two soluble reactants			
Different precipitates have different colours			
Neutralisation is a chemical reaction which happens when an acid and an alkali react together			
The pH of a solution can be measured using a pH meter, and estimated using Universal Indicator and a pH colour chart			

You should be able to:	**I can't do this yet**	**I need more practice**	**I can do this by myself**
Describe the strengths and limitations of a model			
Consider hazards and how to reduce them, in order to carry out practical work safely			
Make predictions of likely outcomes for a scientific enquiry based on scientific knowledge and understanding			
Choose which variables to change, control and measure			
Choose the correct equipment to collect evidence and answer a scientific question			
Evaluate whether observations have been repeated sufficiently to be reliable			

If you have ticked 'I don't understand this yet' or 'I can't do this yet' or mostly 'I need more practice', have another look at the relevant pages in the Student's Book. Then make sure you have completed all the questions in this Workbook chapter and the review questions in the Student's Book.
If you have already completed all the questions, ask your teacher for help and suggestions on how to progress.

Teacher's comments

End-of-chapter questions

1. This question is about chemical reactions with oxygen.

a What compound is made when iron reacts with oxygen?

Tick **one** box.

☐ Iron oxygen ☐ Iron oxide

☐ Rust ☐ Iron hydroxide

b When carbon burns in oxygen gas, carbon dioxide gas is produced.

(i) Write the word equation for the reaction:

__________ + __________ → __________

(ii) Describe a test to show that carbon dioxide had been produced.

____

____

(iii) Explain why the mass of carbon appears to decrease during the reaction.

____

____

c When magnesium reacts with oxygen, a white solid is formed.

(i) Name the white solid. ______________________________

(ii) Use the particle model to predict how the mass of magnesium will change during the reaction. Give a reason for your answer.

____

____

2. When zinc metal is added to hydrochloric acid, bubbles of gas are observed.

Omar wanted to know what the gas was so he carried out some tests. Here are the results.

Test	Observation	Conclusion
Limewater	No change	
Light the gas	Squeaky pop	
Add a glowing splint	The splint went out	

a Complete the table.

Write a conclusion for each test

b Suggest a reason why Omar decided to repeat the tests.

____

3. This question is about precipitate reactions.

a What is a precipitate reaction?

____

____

b Samira is working in the laboratory when she notices that the label has fallen off one of the bottles. She thinks that the bottles contains copper chloride solution. So she carries out a couple of tests to see if she is right. Here are the results.

Test	Observation
Add a few drops of the solution to dilute sodium hydroxide	Blue precipitate
Add a few drops of silver nitrate to the solution	White precipitate

Is Samira correct? Give reasons for your answer.

c When iron chloride reacts with sodium hydroxide a rust coloured precipitate is formed.

(i) Name the precipitate

(ii) A particle diagram can be used to model the reaction.

Complete the diagram:

●● + ●● → ________ + ________

(iii) Evaluate your particle diagram.

4. Milk sample A was left in the kitchen for several days, while sample B was kept in the fridge.

Each day, the pH of both samples was measured using a pH meter.

Here are the results:

Day	pH of milk sample A	pH of milk sample B
Monday	6.7	6.7
Tuesday	6.5	6.7
Wednesday	6.3	6.8
Thursday	5.4	6.7
Friday	4.8	6.5

a Explain why a pH meter was used during this experiment and not Universal Indicator.

b State which sample of milk went sour during the experiment.

c Use data from the table to explain your answer to part (b).

____

____

5. A student added potassium hydroxide, a drop at a time, to a sample of hydrochloric acid containing methyl orange indicator.

She stopped adding the alkali when all the acid had reacted.

a Explain how the student knew when to stop adding the alkali.

____

b State the type of reaction that occurred.

____

c Complete the word equation reaction.

hydrochloric acid + ______________ → ______________ + water

6. Indigestion is caused when stomach acid irritates the lining of the food pipe (gullet).

a Which remedy would you use to cure it?

Tick **one** box only.

☐ Antacid ☐ Milk

☐ Toothpaste ☐ Lemon juice

b Use your knowledge of acids and alkalis to explain why your chosen remedy will work.

____

____

Physics

Chapter 6: Energy

Chapter 7: Forces and their effects

Chapter 8: Sound

Chapter 9: Electricity and circuits

6.1 Energy at work

You will learn:

- To describe the energy transfers that happen in processes and events

1. Which of the following things **must** involve an energy transfer? Tick all that apply.

☐ **a** A change in how an object is moving.

☐ **b** A temperature increase.

☐ **c** A motionless object staying still.

☐ **d** The Sun producing light.

2. When objects are moved they can store energy in two ways.

Complete the sentences using words from the list.

position	**speed**

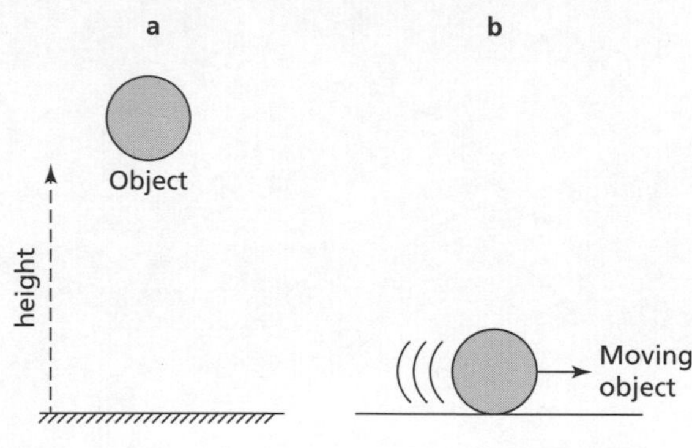

6.1

a An object stores energy because of its ______________________.

b An object stores energy because of its ______________________.

3. The table lists some everyday devices and activities, and the evidence that energy is being transferred. Complete the table.

Device	How device is used	Evidence that energy is being transferred when the device is used
Bunsen burner	Used to heat a liquid in a flask	Temperature of liquid increases
Electric reading lamp	Person uses lamp at night time to see the text in a book	
Electric 'stop' buzzer on a bus	Person presses the button to alert driver they want to get off the bus	
Car engine	Burns fuel to turn wheels	

Table 6.1

4. To make anything change or move, energy is needed. Name the source of energy for each of the following activities.

a A person going for a run.

Show Me

The person uses energy from ____________________.

b A gas heater used to warm a room ____________________.

c An electric lamp in a circuit containing a battery ____________________.

d The Sun producing light ____________________.

5. A scientist says that the food eaten by animals and the fuel used by motor vehicles are very similar. Describe how food and fuel are similar.

____

____

____

Remember

Make sure that you explain this statement and do not just repeat the question. Think about this in terms of energy – for example, when energy is transferred from one object to another, something must happen.

6. In a central heating system in a house, a large tank of water is an energy store.

a Describe how energy is transferred to the tank of water.

____

____

b Use the particle model to explain how the particles of water store energy.

____

____

c Explain why it is important that the tank of water is surrounded by insulating material.

____

____

7. Complete the energy flow diagrams.

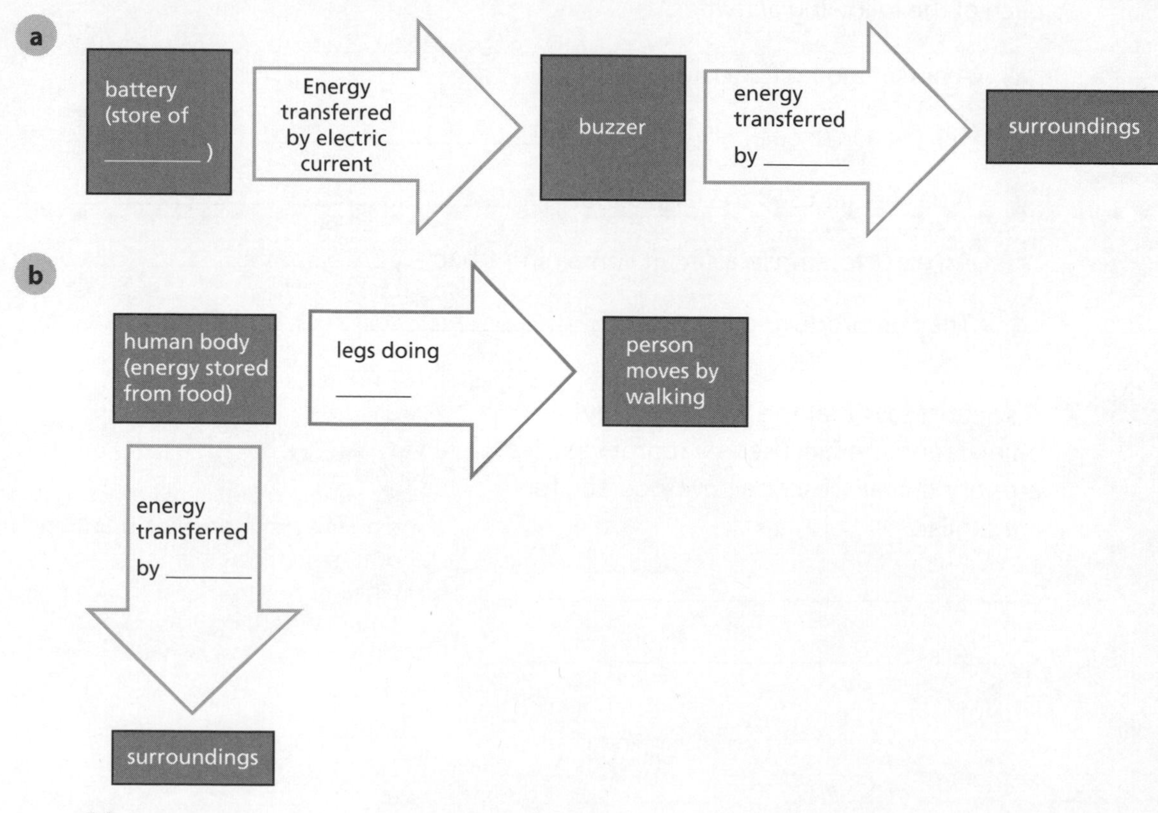

6.2

8. In a shop, a person pushes a shopping trolley to make it move. Scientists say that some of the person's energy is used doing work. Explain what 'work' means in this situation.

____

____

9. Mia holds a ball at head height. She then releases the ball and it falls to the ground.

Challenge

a Describe how the ball stores energy before it is dropped.

____

b Describe how the ball stores energy at the instant before it hits the ground.

____

c Describe the energy transfer that takes place.

____

____

d Explain how Mia could increase the amount of energy stored before the ball is released.

____

6.2 Energy dissipation

You will learn:

- To know that when energy is transferred to make something happen, some energy dissipates and becomes less useful
- To use scientific understanding to predict the likely outcomes of an investigation
- To describe the accuracy of predictions based on results

1. Complete the sentences using words from the list.

wasted	**useful**	**transfers**

Every activity involves ______________________ of energy.

When we operate a device, some energy is used to produce the activity we want.

We call this ______________________ energy.

Some energy is used to produce activities we don't want.

We call this ______________________ energy.

2. The table lists the energy transfers that take place when a cook uses an electric fan oven.

Decide whether the transfer is useful (U) or wasted (W). Write U or W in each blank box. The first one is done for you.

Energy transfer from electrical components …	Useful (U) or wasted (W)
… to fan, to make it turn and circulate hot air.	U
… to outer case of oven, to make it hot.	
… to air and food, to make them hot.	
… to fan, to produce sound as it turns.	
… to surroundings outside oven, to warm them up.	

3. After food has been cooked in an oven, it is hot. We need to take it out of the oven so that it can cool until it is the right temperature to eat.

The food transfers energy to the surroundings. What do we say has happened to the energy that is transferred to the surroundings?

Tick **one** box.

- **a** concentrated ☐
- **b** dissipated ☐
- **c** dissolved ☐
- **d** used ☐

4. An electric screwdriver uses energy to turn screws and push them into objects. We say the screwdriver does useful work.

a Predict **two** ways in which the screwdriver could waste energy.

____

____

Remember

For a **describe** question, you just need to say what happens – in this case what energy transfers occur. You don't need an explanation of why it happens.

b Describe briefly how you could find evidence to test your prediction.

____

____

5. A dynamo (a device that produces electricity) is used to power a bicycle lamp.

Movement from the cyclist turning the pedals causes the dynamo to rotate. This produces electricity.

a Describe the useful energy transfer that takes place ____________________

____.

b Suggest **two** wasted energy transfers that take place ____________________

____.

6. **Challenge**

An inventor claims that they have made a device that 'produces 100% useful energy'. The device is a new kind of lamp that can work off mains electricity or a battery.

a Does the inventor's claim seem reasonable? Use your knowledge of energy transfers to support your answer.

____

____

____

____

b Write a brief description of how you could test your argument.

____

____

____

Self-assessment

Tick the column which best describes what you know and what you are able to do.

What you should know:	I don't understand this yet	I need more practice	I understand this
Energy changes make things happen by transferring energy from a source to other objects or devices			
A moving object stores energy, which we call kinetic energy			
Elastic objects can store energy when they are stretched or squashed			
Objects also store energy when they are lifted higher			
A moving object stores energy, which we call kinetic energy			
Scientists say that energy dissipates. This means that with every useful energy transfer there is always some wasted energy			
The wasted energy becomes more spread out in the environment and so is less useful for making something else happen			
You should be able to:	**I can't do this yet**	**I need more practice**	**I can do this by myself**
Draw a table to record observations			
Use the particle model to explain the link between energy and temperature			
Use flow diagrams to describe energy transfers			
Use understanding of energy to predict the motion of objects in an investigation			

If you have ticked 'I don't understand this yet' or 'I can't do this yet' or mostly 'I need more practice', have another look at the relevant pages in the Student's Book. Then make sure you have completed all the questions in this Workbook chapter and the review questions in the Student's Book. If you have already completed all the questions, ask your teacher for help and suggestions on how to progress.

Teacher's comments

End-of-chapter questions

1. State whether each of the following sentences describes a useful energy transfer or a wasted energy transfer.

- **a** The brakes on a car slow the car down ______________________.
- **b** A crane lifts a container onto a ship ______________________.
- **c** There is a loud bang when someone shuts a door hard ______________________.

2. The sentences describe the energy transfers when water in a boiling tube is heated in a Bunsen flame. The boxes are **not** in the right order.

Write numbers 1 to 4 in the boxes to show the correct order.

- ☐ Energy transferred from boiling tube to heat up substance.
- ☐ Energy transferred from gas flame to heat up boiling tube.
- ☐ Energy of water vapour dissipated in the surroundings.
- ☐ Energy used to turn liquid into water vapour.

3. A heavy box is being pushed up a smooth slope.

6.3

a Name the energy source for the person pushing the box ________________.

b Complete the sentences.

The force to move the box is provided by the ________________________.

When a force moves, we call this transfer of energy doing ________________.

c Describe **two** ways in which energy is being stored by the box.

____

____

d Suggest **one** way in which energy is being wasted.

____

4. Yuri and Priya investigated three different types of lightbulb. Table 6.2 shows their findings.

Practical

Type of lightbulb	Amount of useful energy (%)	Amount of wasted energy (%)	Usual lifetime of lightbulb in years
Filament lamp	10	90	Between 1 and 2
Fluorescent lamp	20	80	Between 2 and 7
LED	75	25	Between 9 and 22

Table 6.2

a Describe the useful energy transfer for all lightbulbs ________________

____.

b Predict and describe the main wasted energy transfer ________________

____.

c Which is the lightbulb that produces the most useful energy compared

to the amount of wasted energy? ______________________________

Challenge

d Use the information in the table to suggest what type of lightbulb should be used in streetlights. Justify your answer.

____

____

____

7.1 Gravity

You will learn:

- To describe the force of attraction between two objects as gravity
- To describe how the masses of two objects affects the force between them

1. Complete the sentences using words from the list.

gravity	**touching**	**force**

We can pull or push an object. When we do this, we are making a ____________________ act on the object.

A force from one object can act on another object even if the objects are not

____________________.

For example, the Sun pulls on the Earth across empty space because of

____________________.

2. We can show forces using arrows. The length of the arrow shows the size of the force.

Complete the diagram by answering the questions.

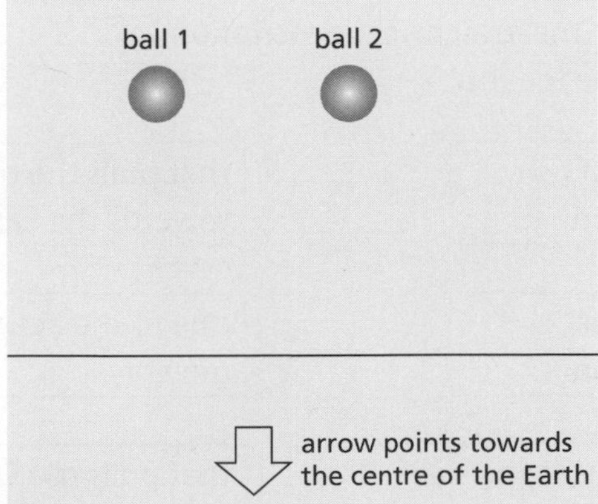

7.1

a On the diagram, figure 7.1, draw an arrow showing the weight of the first ball.

b Explain what the direction of the arrow you have drawn tells us.

____

c The two balls have the same mass. Draw an arrow showing the weight of the second ball.

3. Complete the crossword by answering each clue. The first clue has been answered for you.

(a)
(b)
(c)
(d)
(e)
(f)
A
T
T
R
A
C
T
S

Show Me

a The force of gravity always *attracts*. ✓

b The unit of force. (6 letters)

c The planet we are pulled towards all the time by gravity. (5 letters)

d The reason objects with mass have weight. (7 letters)

e Push or pull on an object. (5 letters)

f Property of all objects that causes the force of gravity. (4 letters)

4. Draw lines between the sentence fragments to make a correct description of gravity and weight. The first sentence is done for you.

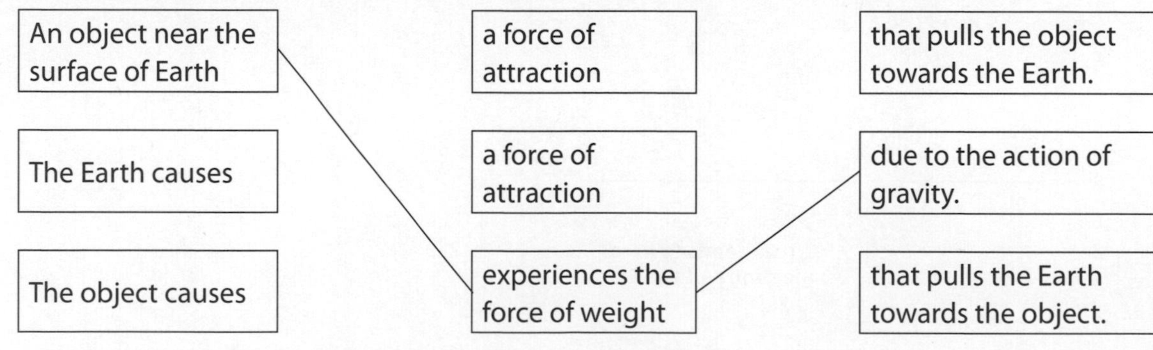
An object near the surface of Earth
The Earth causes
The object causes
a force of attraction
a force of attraction
experiences the force of weight
that pulls the object towards the Earth.
due to the action of gravity.
that pulls the Earth towards the object.

5. Complete the sentences using words from the word list. You may need to use each word once or more than once.

mass	**weight**

The amount of matter an object is made from is called its ______________________.

The force of gravity acting on an object causes the object to have ______________________.

Compared with on Earth, an object on the Moon has the same ______________________

and a smaller ______________________.

6. The diagrams in figure 7.2 show four different ways of showing the forces due to gravity between the Earth and the Moon.

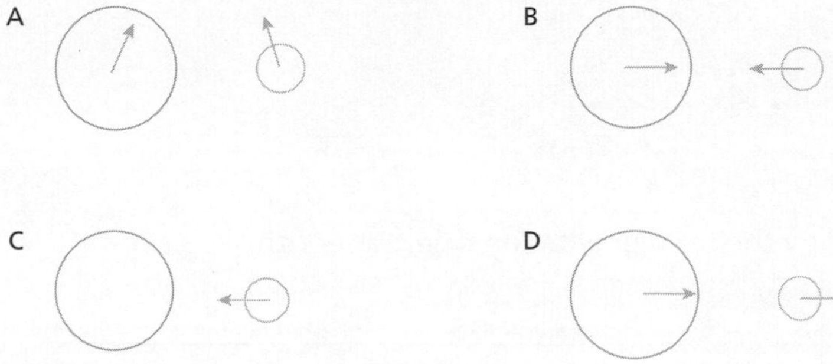

A B

C D

7.2

a Which diagram is correct? ______________________

Show Me

b Explain briefly why each of the other diagrams is incorrect.

1. Diagram ______________________ is incorrect because the force on each object should be towards ______________________.

2. Diagram ______________________ is incorrect because ______________________

______________________.

3. Diagram ______________________ is incorrect because ______________________

______________________.

7. Decide whether each of the following sentences is true (**T**) or false (**F**). Write **T** or **F** in each box. The first sentence has been done for you.

- [T] Near the Earth, all objects that have mass experience the force of gravity.
- [ ] Gravity is a force that acts at a distance.
- [ ] All objects that have mass repel each other.
- [ ] Between two objects, the force of gravity only acts on the smaller object.

8. The table below shows the values of approximate gravitational field strength for Uranus, Mars and Earth.

Planet	Approximate gravitational field strength (N/kg)
Uranus	9.0
Mars	4
Earth	10

Challenge

a Explain what is meant by the term gravitational field strength.

____

b Suggest why the gravitational field strength for Earth is larger than that of Mars.

____

c Calculate the weight of a 78kg person on Earth and Uranus.

____

____

d Name the piece of equipment used to measure the weight of an object.

____

e Explain how and why the weight of a person on Earth is different from their weight on the Moon.

____

____

7.2 Air resistance

You will learn:

- To describe how air resistance affects the speed of moving objects
- To understand why movement is opposed in a vacuum

1. Complete the sentences using terms from the list.

air resistance	gravity	friction

The force of attraction between all objects with mass is due to ____________________.

Two solid objects rubbing against each other experience the force of ____________________.

An object moving through the atmosphere is slowed down by the force due to

____________________.

Read the descriptions of an experiment and answer the questions that follow.

A scientist uses the equipment in the diagram to investigate how objects fall, including whether air resistance has an effect.

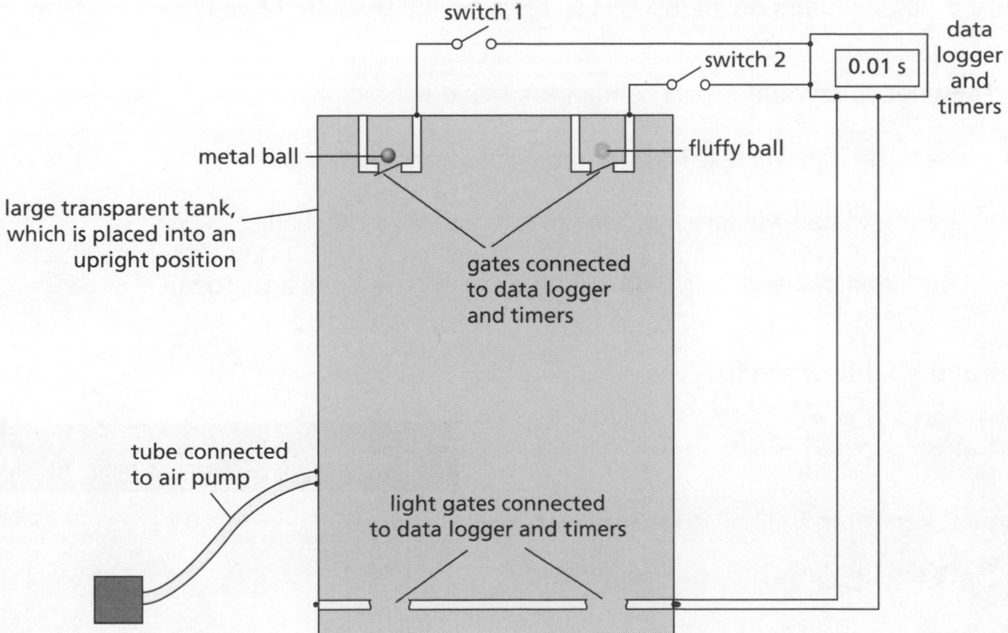

7.3

She uses a small fluffy ball and a small metal ball as her test objects.

Data loggers are electronic machines that can be used to collect and record data, such as time or current, during an investigation. In this investigation the data loggers are being used to record the time taken for the metal ball and the fluffy ball to drop down through the tank.

Light gates are sensors that can detect the movement of an object through them. The light gates being used in this investigation detect the movement of the metal ball and the fluffy ball as they drop through them.

The air pump can be used to remove air from the tank. This is not done in the first test.

2. Practical

A good hypothesis predicts how measurements of quantities will change in an investigation, and is testable. Choose the best hypothesis for this investigation. Tick **one** box.

a Different objects fall in different ways. ☐

b All objects fall at the same rate, except for flat objects with low mass. ☐

c All objects fall at the same rate in a vacuum, but some objects are slowed down more than others due to air resistance. ☐

d All objects fall. ☐

For the first test, the scientist checks that the two objects are held at the same height in the air. She releases the objects by pressing two separate switches that open the gates. Switch 1 starts both timers. The light gate for each ball stops each timer. The air pump is not switched on for this test so the air is not being removed from the tank.

3. Practical

Predict what you would expect to happen. Tick **one** box.

a The fluffy ball will take less time to fall than the metal ball. ☐

b The metal ball will take less time to fall than the fluffy ball. ☐

c The metal ball and fluffy ball will take exactly the same time to fall. ☐

4. Practical

Look at the table of results.

	Test 1	Test 2	Test 3
Time taken for metal ball to fall in seconds	0.32	0.32	0.31
Time taken for fluffy ball to fall in seconds	0.41	0.45	0.43

Table 7.1

a Suggest why the results are different for each repeat of the test.

____

____

b Suggest a way to improve the accuracy of the results.

____

c Explain the difference between the time taken for the metal ball to fall and the time taken for the fluffy ball to fall.

____

____

5. Practical

Describe how you could use the equipment shown in the diagram to investigate the time taken for the metal ball and the fluffy ball to fall with no air resistance.

____

____

____

6. Practical — Show Me

Predict what you would expect to happen if there was no air resistance.

No air resistance means there is no force acting to ______________________.

This means that the fluffy ball and the metal ball should fall ______________________.

7. Air resistance is a force that acts when an object moves through the air.

Read each first half-sentence (a) to (d). Then match each second-half sentence to the correct first-half sentence. One line has been drawn for you.

(a) A skydiver falling through the air	**(1)** uses streamlining to reduce the air resistance
(b) An spacecraft carrying people	**(2)** does not need to worry about air resistance, as there is no air.
(c) An aeroplane manufacturer	**(3)** uses the extra air resistance of a parachute to slow her fall.
(d) An astronaut on the Moon	**(4)** uses heat-resistant materials to protect passengers from the heat produced by air resistance on re-entering the atmosphere.

8. **Challenge** When NASA developed the parts of the rocket and landing craft to take astronauts to the Moon and back, they needed to keep the astronauts safe at all stages of the journey. However, they also needed to reduce the mass of each part of the rocket as much as possible.

a Suggest why it was important to keep the mass as low as possible.

____

____

b One way of reducing mass was to only use heat-resistant materials where they were needed. Explain why there were heat-resistant materials on the outside of the capsule in which the astronauts sat for launch and for returning to Earth, but not on the craft used to land on the Moon.

____

____

____

Self-assessment

Tick the column which best describes what you know and what you are able to do.

What you should know:	I don't understand this yet	I need more practice	I understand this
The weight of an object on Earth is the pull force of the Earth's gravity on it			
Weight is measured in newtons			
The size of the force of gravity between two objects depends on the masses of the objects			
Air resistance tends to slow down moving objects			
There is no air resistance to resist movement in a vacuum			
You should be able to:	**I can't do this yet**	**I need more practice**	**I can do this by myself**
Identify whether a hypothesis is testable			
Interpret results in a table			

Make predictions and review them against evidence			
Make predictions referring to previous scientific knowledge and understanding			
Explain why accuracy is important			
Evaluate investigations and suggest improvements			

If you have ticked 'I don't understand this yet' or 'I can't do this yet' or mostly 'I need more practice', have another look at the relevant pages in the Student's Book. Then make sure you have completed all the questions in this Workbook chapter and the review questions in the Student's Book. If you have already completed all the questions ask your teacher for help and suggestions on how to progress.

Teacher's comments

End-of-chapter questions

1. I can be a push or a pull. I can be of many different types. I am always there if an object speeds up, slows down or changes direction. I am invisible but I can be measured. Objects don't always have to be touching for me to appear.

What am I? ______________________

2. Explain why it is more difficult to lift a box full of objects than it is to lift the same box when it is empty.

____

____

____

3. The diagram shows a box of medical supplies being dropped from an aircraft.

a Complete the diagram showing the forces acting on the box as it falls.

b A parachute opens. Complete the diagram to show the new forces acting on the box.

7.4

c Explain how a parachute helps to prevent the medical supplies being damaged when they reach the ground.

____

____

____

7.5

4. The weight of an object near the planet Jupiter is 2.5 times greater than the weight of the same object near Earth.

a Name the force that causes an object to have weight. ____________________

b Deduce which planet has more mass: Earth or Jupiter? ____________________

Scientists who design space probes to investigate Jupiter have to take the increased weight into account.

c Which row of the table shows the correct information about the weight of the space probe? Tick **one** box.

	Direction of the weight of the space probe	Effect of the space probe on Jupiter
☐ **A**	Towards the centre of Jupiter	No effect
☐ **B**	Towards the centre of Jupiter	A force towards the centre of the space probe
☐ **C**	Away from the centre of Jupiter	A force towards the centre of Jupiter
☐ **D**	Away from the centre of Jupiter	No effect

5. Mia holds a golf ball and a larger fluffy ball at head height. She then lets go of both balls and they fall to the ground.

Practical

Challenge

The balls will accelerate (get faster) until they reach the ground.

a Explain what causes this to happen.

___

Mia observes differences between how the golf ball and fluffy ball fall.

b Predict what you would expect to observe. Will both balls take the same time to reach the floor?

___

c Explain your answer to part (b).

___

___

d Mia suggests that a similar experiment could be set up inside a large airtight container, and all the air pumped out of the container. Describe what you would expect to observe.

___

___

8.1 How sound travels

You will learn:

- To describe how sound waves are made by the vibration of particles
- To explain why sounds do not move through a vacuum
- To collect and record measurements
- To explain the importance of precision and accuracy

1. Complete the sentences using words from the list.

wave	**force**	**vibrates**	**burns**

Sound is produced when an object or a substance ____________________.

We represent how sound travels as a ____________________.

2. Look at the diagram of a model of a sound wave passing through a substance. Complete the sentences using words from the list.

direction of travel of wave

8.1

particle	**sound**	**wave**	**chemical reactions**	**collisions**
		vibrates	**stays still**	

Each small ball in the diagram represents one ____________________.

The blue arrows on one ball show that each ball ____________________.

The sound wave moves through the substance because of ____________________ between the particles.

3. Look at the diagram figure 8.2. An electric circuit including a buzzer is set up inside a glass tank.

Mia switches on the circuit so the buzzer makes a loud noise. She then switches on a pump that gradually pumps the air out of the tank.

Mia observes that the sound gets quieter until she cannot hear the sound at all.

Give a possible hypothesis for Mia's experiment and explain how observations can be used to test this hypothesis.

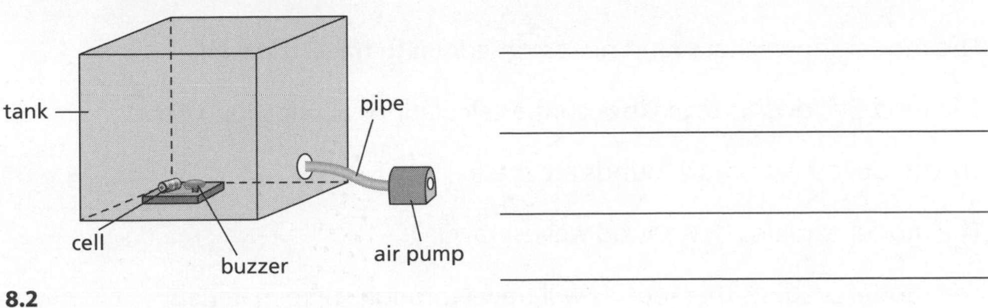

8.2

4. Carlos wants to investigate which materials sounds can travel through.

Practical

He has been given the following materials to use in his investigation.

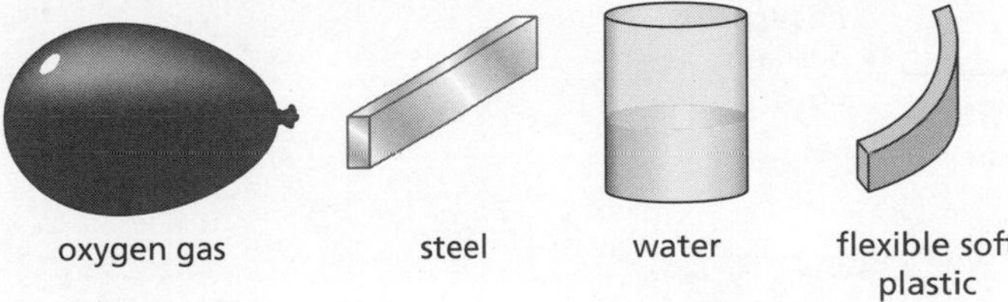

8.3

He has also been given some instructions about the steps he needs to take in planning his investigation. He has written some notes about the things he can choose between.

Help Carlos by matching each of his notes to the instruction. Write the **letter** of **each** statement in the correct place. The first one has been done for you.

Instructions

1 Decide which variable to change. B

2 Decide which variable will be measured as it changes. ☐

3 Decide which variables to keep constant (the same). ☐

4 Note any variables that could affect the results but which you cannot control. ☐

Carlos's notes

A Temperature difference.

B Choice of object.

C Whether sound is detected.

D Background noise (from other people and other rooms).

5. Jamila has made some notes about the strengths of the particle model of sound travelling through materials. Which notes are true and which are false? Write '**T**' for TRUE and '**F**' for FALSE.

☐ The model shows that sound needs a medium to travel through.

☐ The model allows us to make accurate calculations about sound waves.

☐ The model explains how sounds are made.

☐ The model explains how sound waves travel.

☐ The model predicts that sounds will travel through some materials but not others.

6. The diagram shows a metal ruler attached to a table so that half of the ruler sticks out over the edge. Lily finds she can make a sound that she can hear by bending the ruler at the end and releasing it.

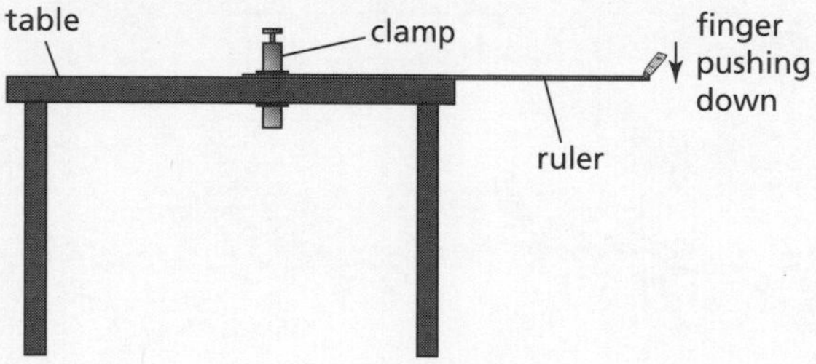

8.4

Describe the steps in the process that cause a sound to travel through air.

Show Me

After the ruler is pushed and released, it vibrates.

The moving ruler causes the air particles near the ruler to ______________________.

These air particles collide with more ______________________.

The movement of the air particles between the ruler and Lily's ear

____

7. Challenge

Professor Green lives near an airport. Every time an aircraft passes over, it is very noisy inside the house. She decides to replace all the windows to reduce the amount of noise.

a The diagrams show four different choices of window designs. Which design, a, b, c or d, will be the best at keeping out noise?

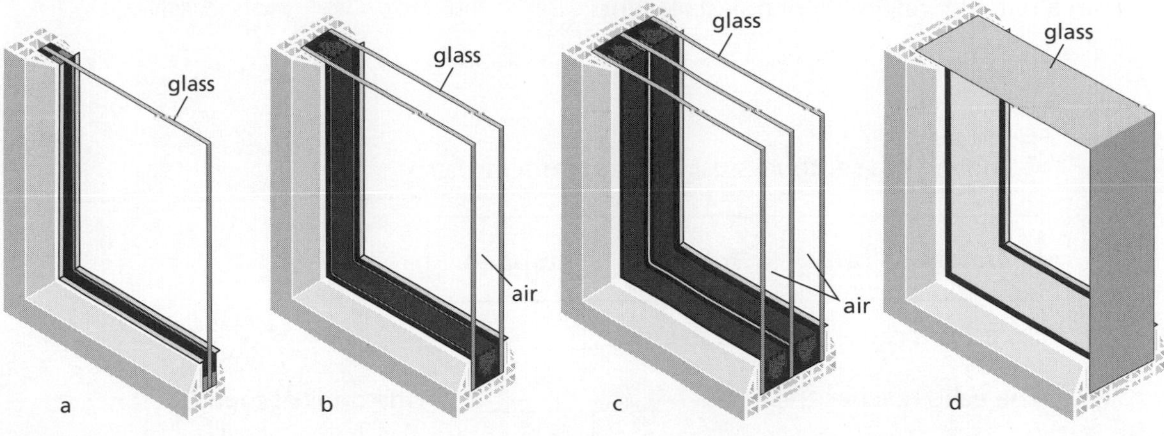

8.5

___

b Explain your choice.

___

c Suggest **one** way in which your chosen design could be made even better at keeping out sounds.

___

8.2 Echoes

You will learn:

- To explain how echoes are made
- To know how to choose and use the correct apparatus
- To evaluate the reliability of measurements

1. Complete the sentences using words from the list.

before	**after**	**reflected**	**stopped**

An echo is caused when a sound is ________________ .

The echo reaches the listener ________________ the original sound.

2. The diagram in figure 8.6 shows a view from above of a building and a road. Simran stands at point S and calls out to Rena. Rena stands at point R.

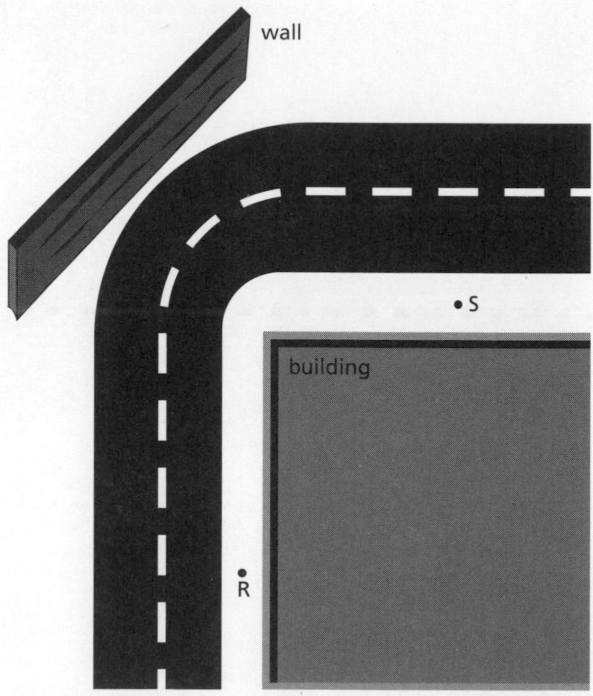

8.6

Explain how Rena can hear Simran, even though she cannot see her.

____

____

3. Describe what happens when we hear an echo.

Show Me

When a sound wave meets a hard surface, it is reflected.

This causes the direction the sound wave travels in to ______________________________

This sound wave reaches our ears later than ______________________________

4. Lily is a scientist investigating a lake. For part of her study, she uses sound echoes to measure the depth of the lake.

Practical

She uses a boat to move in a straight line from one end of the lake to the other. Every 100 m along the line, she makes a measurement.

She uses a waterproof microphone and data logger to measure the time it takes between a sound being made and the echo to return from the bottom of the lake.

Remember

When you need to write a long explanation, plan your answer in a logical order. Write one or two short sentences for each stage of your explanation.

The data logger is a small electronic device with sensors that can be used to collect and record data during an investigation.

In this investigation the data logger is attached to the microphone to record the time taken between the sound being made and the echo being detected.

These are her results.

Position in m	0	100	200	300	400	500	600	700	800
Time to detect echo in s	0.00	0.32	0.85	1.01	1.24	1.32	1.13	0.64	0.00

Table 8.1

a Use the data to make a conclusion about how the water depth changes along this line.

____

____

____

____

____

Lily repeated the measurements using the same line across the lake. In the second set of data the results included times to detect the echo of 1.19 s at 400 m and 1.30 s at 500 m.

b Does Lily have enough results to be confident about calculating the depth of the lake? Explain your answer.

____

____

5. The diagram shows an unusually shaped building.

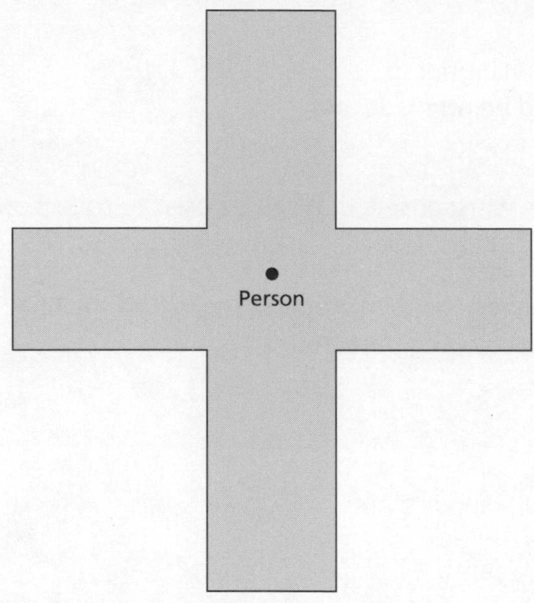

8.6

A person standing at the centre of the building makes a noise. They hear four echoes of the noise.

Explain what is happening.

____

____

____

____

6. Ultrasound describes a type of sound wave. These sound waves are used in medical devices to check the health of babies before they are born.

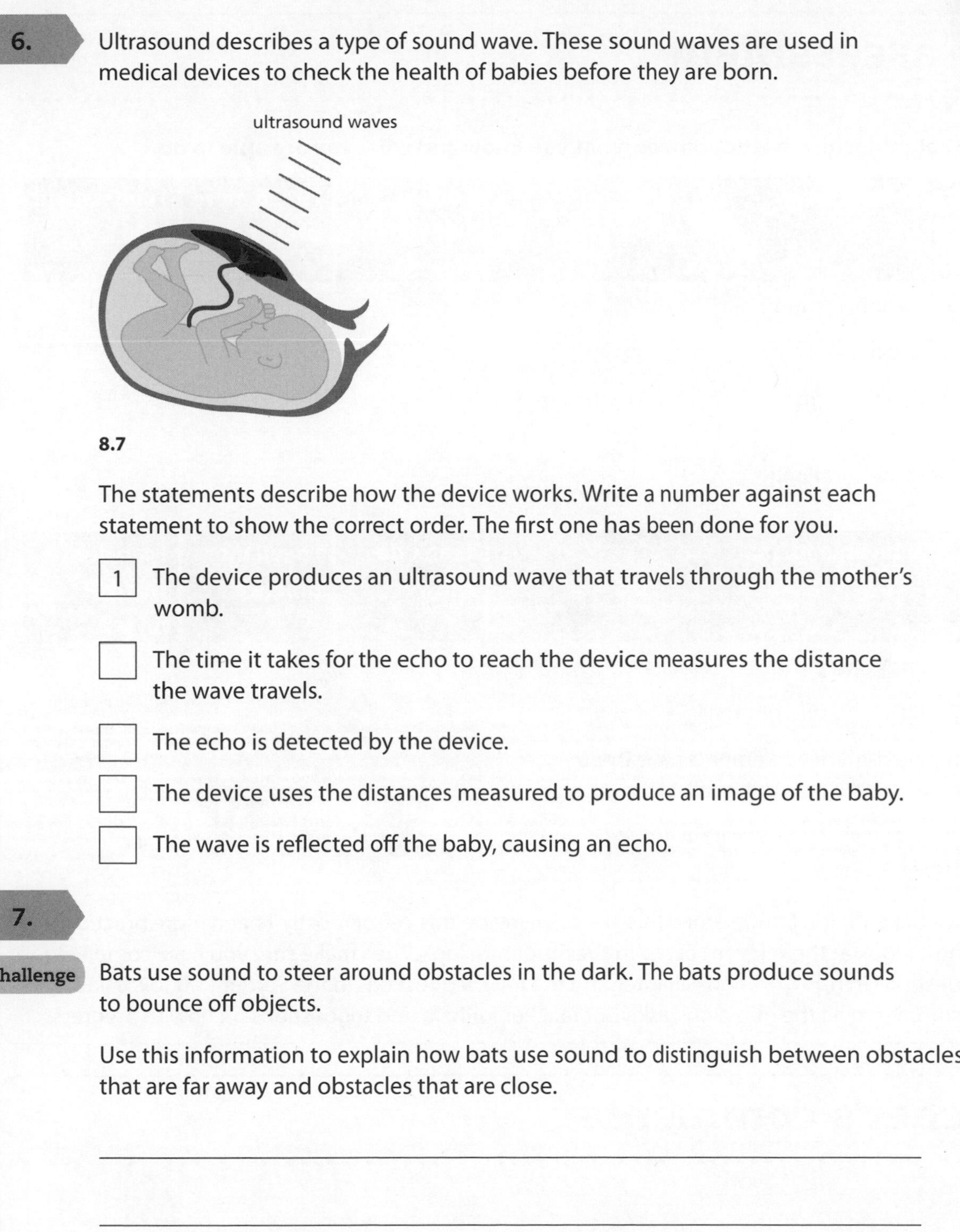

8.7

The statements describe how the device works. Write a number against each statement to show the correct order. The first one has been done for you.

- [1] The device produces an ultrasound wave that travels through the mother's womb.
- [ ] The time it takes for the echo to reach the device measures the distance the wave travels.
- [ ] The echo is detected by the device.
- [ ] The device uses the distances measured to produce an image of the baby.
- [ ] The wave is reflected off the baby, causing an echo.

7.

Challenge Bats use sound to steer around obstacles in the dark. The bats produce sounds to bounce off objects.

Use this information to explain how bats use sound to distinguish between obstacles that are far away and obstacles that are close.

____

____

____

____

Self-assessment

Tick the column which best describes what you know and what you are able to do.

What you should know:	I don't understand this yet	I need more practice	I understand this
Vibrations produce sound			
Sound travels as a wave through a medium			
Sound travels through the air by making the air particles vibrate			
Echoes are sound reflections			

You should be able to:	I can't do this yet	I need more practice	I can do this by myself
Identify trends and patterns in results			
Explain how you would get reliable results			
Evaluate whether measurements have been repeated enough times to be reliable			
Describe how well a model explains how sound waves travel through air			

If you have ticked 'I don't understand this yet' or 'I can't do this yet' or mostly 'I need more practice', have another look at the relevant pages in the Student's Book. Then make sure you have completed all the questions in this Workbook chapter and the review questions in the Student's Book. If you have already completed all the questions, ask your teacher for help and suggestions on how to progress.

Teacher's comments

End-of-chapter questions

1. Some scientific measuring devices are very delicate. They need to be kept inside a container and protected from all vibrations.

What should be in the container to stop sounds affecting these devices? Tick one box.

☐ Vacuum

☐ Air

☐ A liquid such as oil

2. The diagram, figure 8.8, shows a line of air particles.

8.8

a Draw a diagram to show the effect of a sound wave passing along this line of particles.

b A hard surface is placed at one end of the line of particles, at right angles to the line. Describe what happens to the sound wave when it reaches the surface.

____

____

3. Colin investigates echoes using the equipment in the diagram, see figure 8.9.

Practical

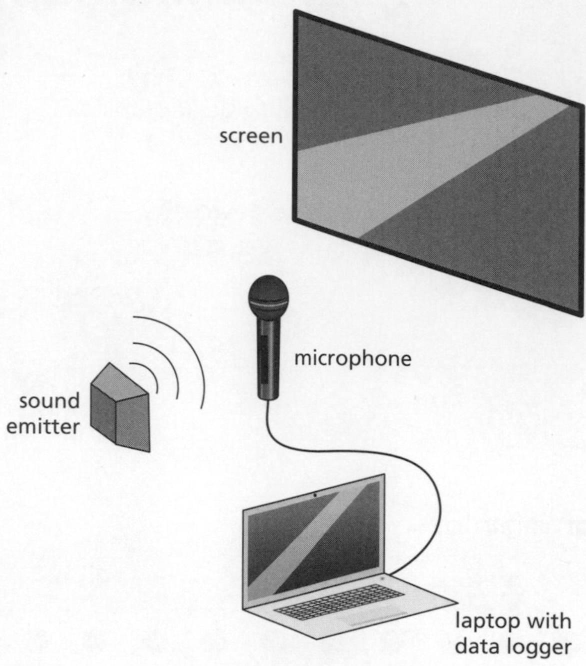

8.9

The sound emitter makes a single, short, loud noise. The microphone detects the sound from the emitter. A short time later, the microphone detects the echo from the screen. The data logger is a small electronic device with sensors that can be used to collect and record data during an investigation.

In this investigation the data logger is attached to the microphone to record the time taken between the sound being made and the echo being detected.

Colin writes this hypothesis for his investigation:

If the distance from the microphone to the screen is increased, the time between the sound and its echo arriving at the microphone will increase.

a Is Colin's hypothesis a good one? Explain your answer.

____

____

The table shows Colin's results.

Distance in m	1.0	2.0	4.0	6.0	10.0
Time in s	0.006	0.010	0.025	0.033	0.060

Table 8.2

b Explain whether (or not) Colin's hypothesis is supported by these results.

____

____

c Suggest how Colin could make his results more reliable.

____

____

4. Railway engineers can use a microphone and recording equipment to detect the vibrations of a train moving on the rails. The equipment detects an approaching train for many seconds before the engineers can hear the sound of the train.

Challenge Use your knowledge of sound waves to explain what the engineers observe.

____

____

____

____

9.1 Charge flow in circuits

You will learn:

- To describe electricity using a simple model
- To investigate which materials conduct electricity and which do not

1. What do we call the movement of electrons through a conductor? Tick **one** box.

☐ Electric current

☐ Electric force

☐ Magnetic force

☐ Electric insulation

2. Complete the following sentences using the words from the box.

protons	**neutrons**	**electrons**	**flow**	**roll**

Electricity involves the movement of charged particles called ________________ .

These charged particles __________ around a circuit.

3. Electric charge can move around circuit. For this to happen, we say the circuit must be ______. Tick **one** box.

☐ Hot

☐ Broken

☐ Dry

☐ Complete

4. Each of the following statements describes properties of materials. Choose whether each statement describes an electrical conductor or an electrical insulator. Write 'C' for a conductor and 'I' for an insulator.

☐ Allows electrically charged particles to pass through easily.

☐ Does not allow electrically charged particles to pass through easily.

☐ Is made from plastic and used as the outer covering for wires in electrical circuits.

☐ Is made from metal and used in wires in electrical circuits.

Questions **5** to **8** are about a model of electric circuits. Read the description. Then answer the questions.

We can make a model of the flow of electric charge in a circuit by asking a class of students to sit in a circle. Each student has one other student on their right, and one on their left.

Each student holds hands with the students on either side of them. The students and their arms make a big circle.

One student is the cell. They squeeze the hand of the student on their left. Each student in turn 'passes on' the squeezed hand until it has moved all the way around the circle.

5. What does the squeezed hand passing round the circle represent in an electric circuit?

____

6. Describe how the students could show an open switch that breaks the circuit.

____

7. Suggest how you could show a component called a buzzer using this model.

____

8. Two strengths of this model are that it is easy to set up, and that it models the need for a circuit to be complete very well.

Suggest a limitation of this model.

Show Me

A real circuit moves many millions of electrons but ____________________

____

9. **Challenge** The diagram shows a slice through a piece of wire used to connect a buzzer used in a house into a circuit.

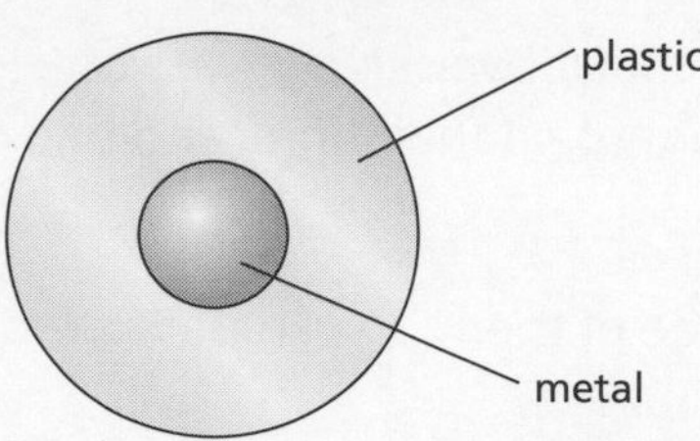

a Explain why the metal at the centre of the wire is needed.

____

____

b Suggest a reason why the metal has been put inside the plastic casing.

____

____

9.2 Circuit diagrams

You will learn:

- To represent circuits using diagrams and conventional symbols
- To investigate the effect of changing the numbers of cells and lamps in a circuit

1. Name each of the components in the diagrams below.

Choose your answers from this list.

switch	cell	battery	ammeter	buzzer	lamp

______________ ______________

______________ ______________

____

2. Draw the symbol for an ammeter.

3. Describe what the current in a circuit represents.

Show Me

The current represents a ______________________________.

4. Look at the diagrams below. For each diagram, complete the circuit so that all the components are complete and connected. For **circuit a**, use just *one straight* connecting wire. For **circuits b**, **c** and **d** complete the components.
Note that you do not need to draw the switches as closed.

a

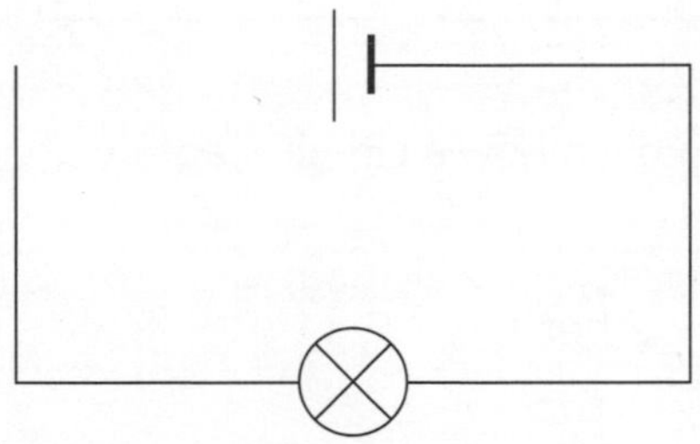

b

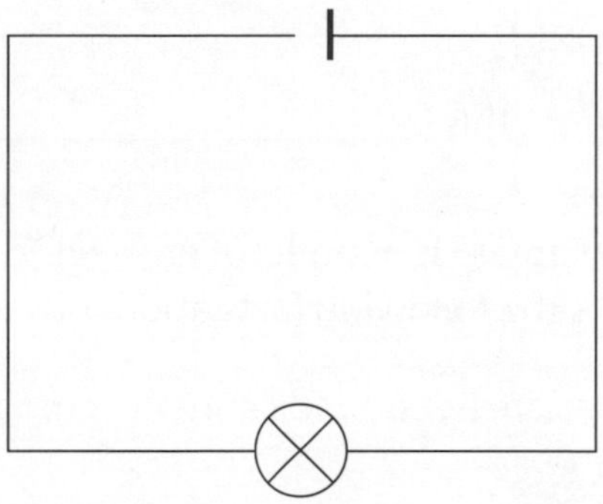

c

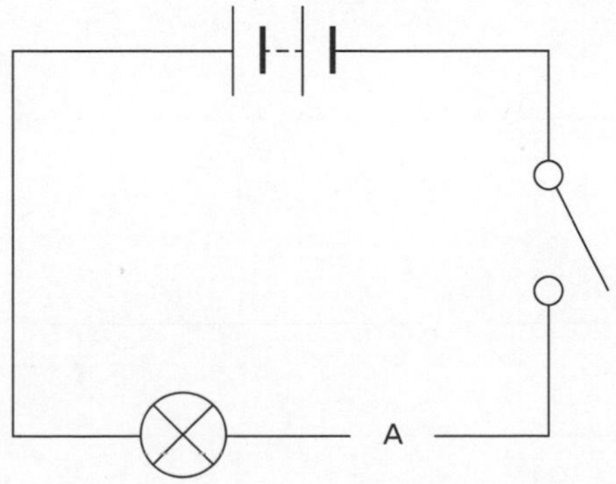

A

d

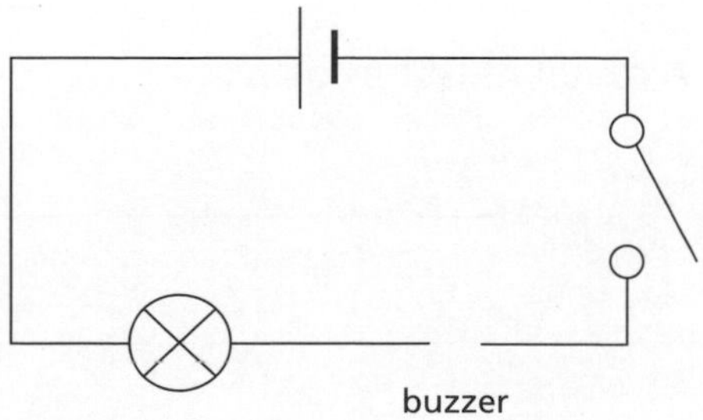

buzzer

5. The diagram of a circuit contains three mistakes. Describe the mistakes and how to correct them.

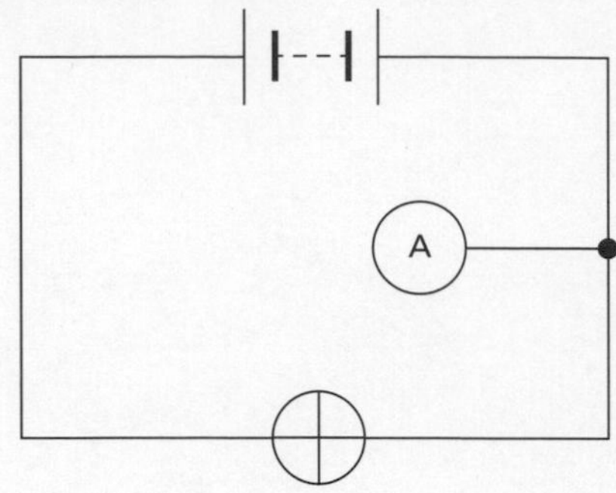

1. ______________________________.

Correct this by ______________________________.

2. ______________________________.

Correct this by ______________________________.

3. ______________________________.

Correct this by ______________________________.

6. A circuit diagram is a model of an electric circuit. Explain how a circuit diagram represents the following facts about electric circuits.

a Wires are needed to connect components together.

____

b A battery is a component that contains more than one cell.

____

c A circuit must be complete.

____

____

9.3 Currents in series circuits

You will learn:

- To know how current can be measured in a series circuit
- To describe how current in a series circuit is affected by adding cells and lamps

1. Which of these components can be used to measure electric current? Tick **one** box.

☐ Battery

☐ Ammeter

☐ Buzzer

☐ Switch

2. Explain why every circuit must be complete for it to work.

Challenge

Current is a flow of ______________________________.

It must be able to flow from one ______________________________.

If a circuit is not complete the current ______________________________.

3. What does a cell or battery provide usefully when it is connected in a circuit? Tick **one** box.

☐ It makes new charges.

☐ It produces heat energy that is transmitted along the wires and warms the circuit components.

☐ It transfers chemical energy to electrical energy to 'push' the current.

☐ It stores kinetic energy of the charges in the circuit.

4. Look at circuits a and b.

a

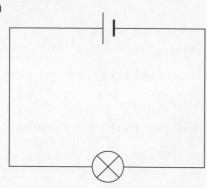

b

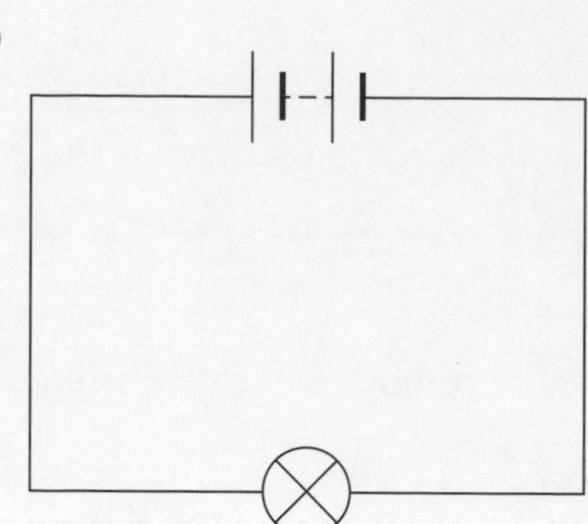

In which circuit, a or b, will a larger current flow? ____________________

5. Describe how you could measure the current in a circuit.

____

____

Questions **6** to **8** are about circuits and components. Look at the diagrams below which show two different circuits. Then answer the questions.

a

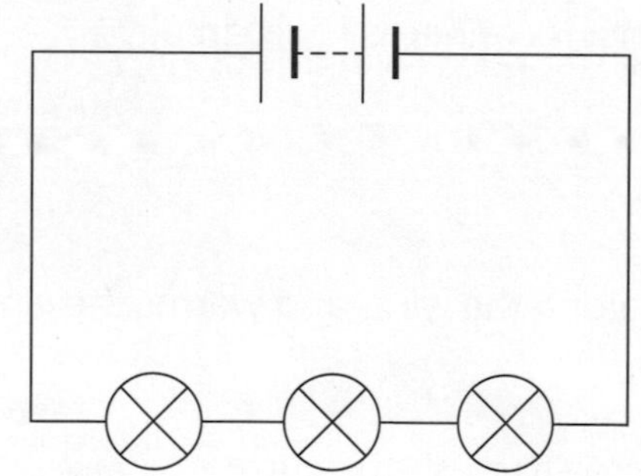

b

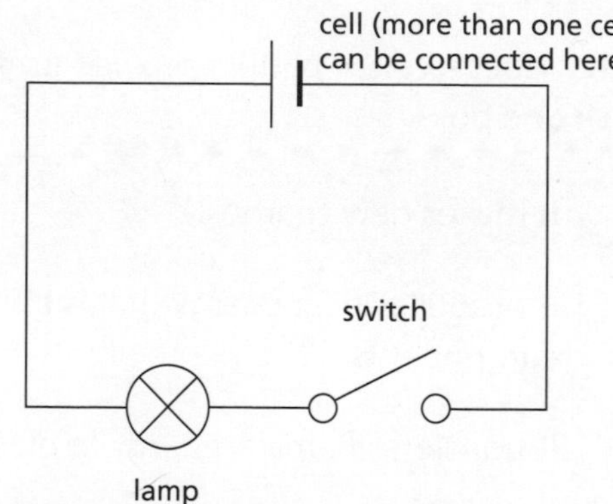

6. Priya builds **circuit a**. She finds that the lamps do **not** shine brightly.

a Suggest what Priya could add to the circuit to increase the current and make the lamps shine more brightly.

____

b Use your knowledge of energy and energy transfers to explain why the lamps will shine more brightly when this component is added.

____

____

____

____

7. Look at the diagrams of circuits a and b.

Show Me

a Compare the circuits. (Assume the number of components in circuit b is kept as shown in the diagram.)

*Circuit **a** contains two cells,* ______________________________ .

*Circuit **b** contains one cell,* ______________________________ .

b Predict whether the lamp in circuit **b** will shine more, less or equally brightly than each of the three lamps in circuit **a**.

____

8. Chen builds a circuit to measure the size of the current when different numbers of components are used. He draws a diagram of his circuit (**circuit b**). More than one lamp can be connected in the circuit. More than one cell can be connected in the circuit.

Practical

Challenge

Here are Chen's results:

Number of lamps used	Number of cells used	Current in amperes
1	1	0.6
2	1	0.3
1	2	1.2
2	2	?

a Chen has made a mistake in his diagram. Which component has he left out of his diagram?

____

b Here is a copy of the diagram. Add the missing component to the circuit.

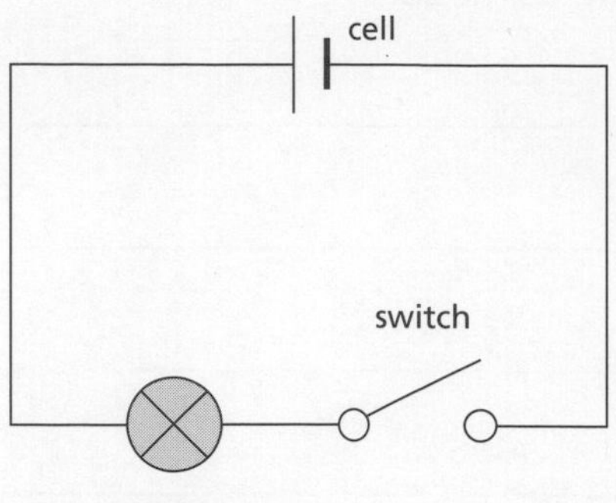

c Predict the missing value for the current in the final row of the table.

___

d Explain why you chose the value you gave in part **c**.

___

___

___

Self-assessment

Tick the column which best describes what you know and what you are able to do.

What you should know:	I don't understand this yet	I need more practice	I understand this
Current is a flow of charge in a circuit			
Cells provide energy to make current flow			
Conductors are materials which allow current to flow			
Insulators are materials which do not let current flow			

Metals are good conductors			
Components in a circuit can be represented by symbols			
Changing the components in a circuit can change the current			
Current is measured in amps using an ammeter			
The current is the same all round a series circuit			
Adding cells to a circuit increases the current			
Adding extra lamps to a circuit decreases the current			

You should be able to:	**I can't do this yet**	**I need more practice**	**I can do this by myself**
Use a model to describe the flow of electricity			
Construct a series circuit			
Explain experimental results using a scientific model			
Construct and compare circuits from circuit diagrams			
Use an ammeter to measure current			
Describe strengths and limitations of models of electricity			
Decide which apparatus to use			
Use explanations to make predictions			

If you have ticked 'I don't understand this yet' or 'I can't do this yet' or mostly 'I need more practice', have another look at the relevant pages in the Student's Book. Then make sure you have completed all the questions in this Workbook chapter and the review questions in the Student's Book. If you have already completed all the questions ask your teacher for help and suggestions on how to progress.

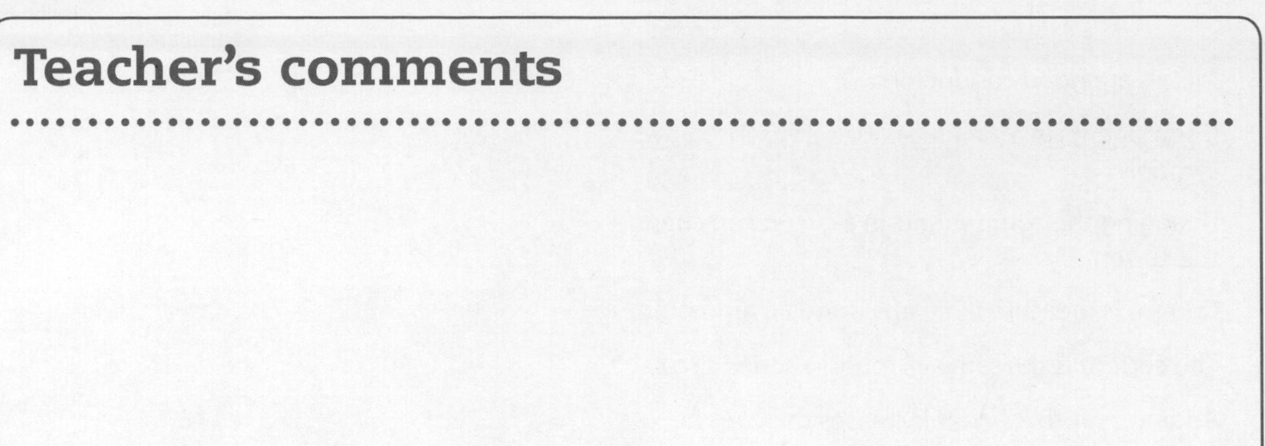

End-of-chapter questions

1. The table shows the current measured in samples of different materials when each is connected in a test circuit.

Material	Current in A
copper (a metal)	1.5
glass	0.0
plastic	0.0
aluminium	1.2

a Which materials are conductors?

b Draw a diagram of a circuit that could be used to do these tests. The sample has been placed for you. Check that you include all the components you need. Include a switch.

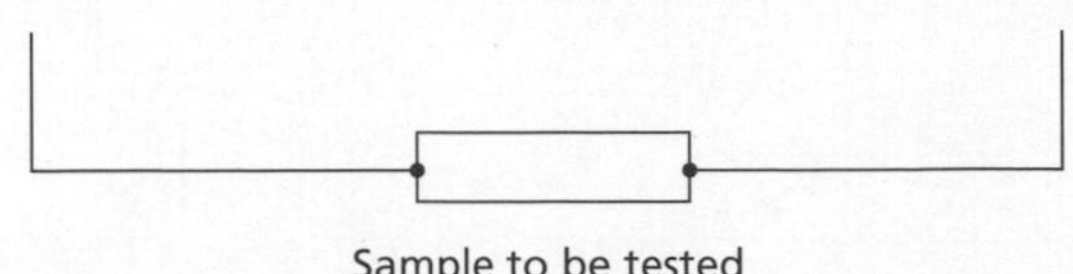

Sample to be tested

2. The diagrams show four different circuits **a** to **d**.

a

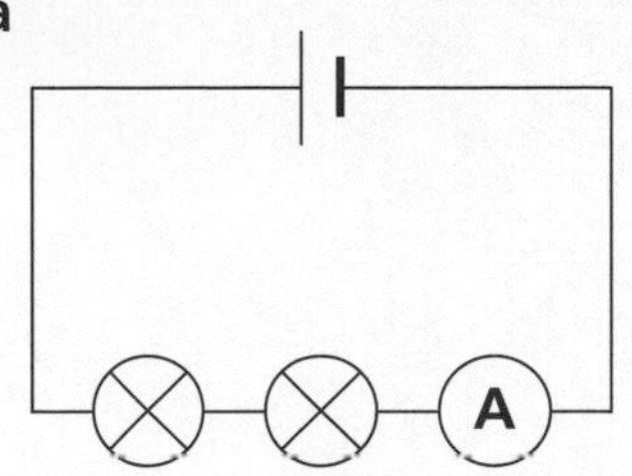

b

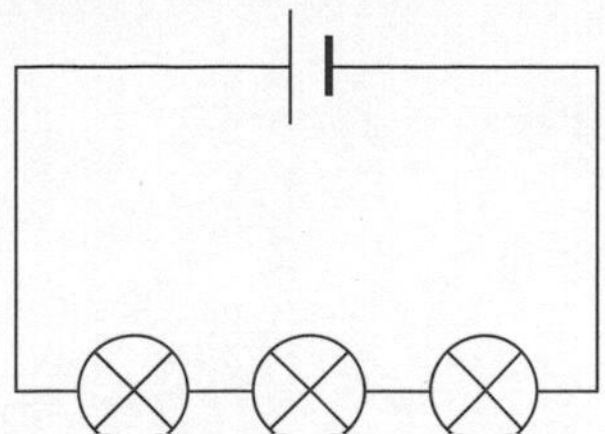

c

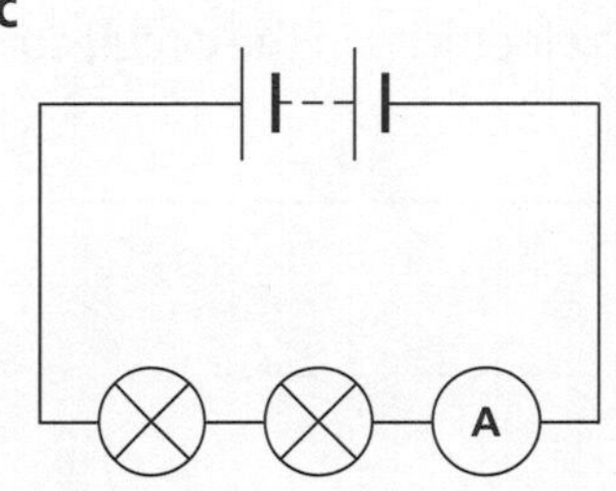

d

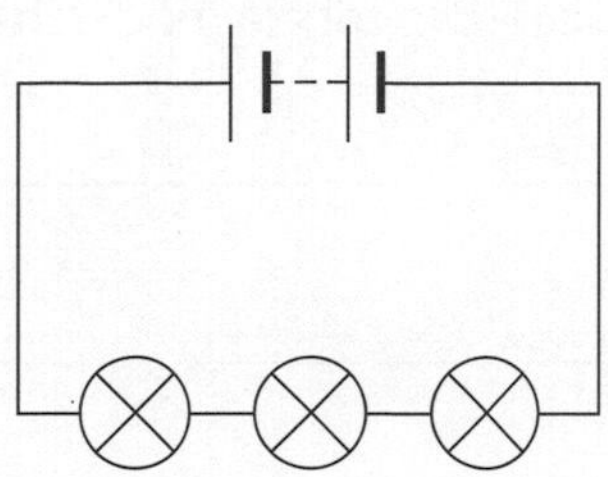

a Which circuits include a component for measuring current?

____

b Name the component used to measure current.

____

c In which circuit would you expect the lamps to shine brightest? Explain your answer.

____

____

d Draw the symbol for a buzzer.

e Draw a diagram of a circuit used for a burglar alarm. It should contain a battery, a lamp, a buzzer and a switch. The lamp and buzzer should both work when the switch is closed.

3. The graphs show the amount of electric charge flowing through two different materials, X and Y, when they are placed in electric circuits.

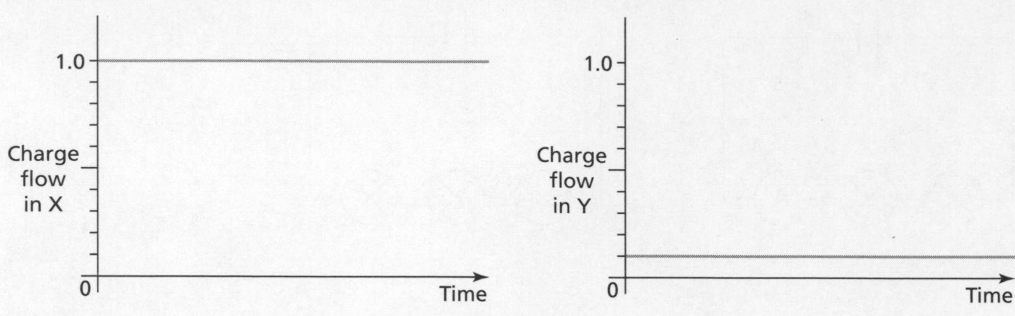

a Material X is a conductor. Describe a model that explains electricity in the circuit containing X.

____

____

____

____

b Suggest whether Y is a conductor or an insulator. Explain your answer.

____

____

____

____

4. Practical

Angelique investigates the current in a circuit, and how it changes when the number of lamps used is changed. Look at the circuit diagram.

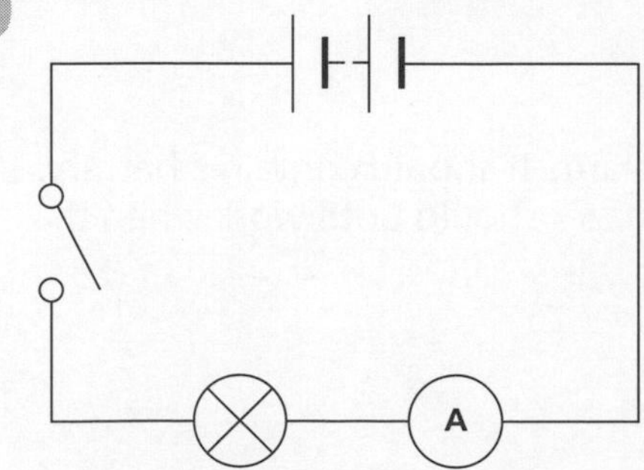

The diagram shows a single lamp, but Angelique can add more lamps in that position.

a Predict what you would expect to observe when the number of lamps is increased.

Look at the table of Angelique's results.

Number of lamps	Current in A
1	2.0
2	1.8
3	0.7
4	0.5

b Do these results fit with your predicted observations? Describe how the results do (or do not) fit your prediction.

c Which, if any, numbers of lamps should Angelique test again? Explain your answer.

5. **a** Explain what an electrical current is.

b Describe how you can increase the amount of current in a circuit without taking any components out of the circuit.

Earth and Space

Chapter 10: The Earth and its atmosphere

Chapter 11: The Earth in space

10.1 The Earth's crust

You will learn:

- To describe the plate tectonics model
- To describe how the model of plate tectonics took time to be accepted by most scientists
- To describe how evidence can be used to test a hypothesis
- To discuss how scientists develop knowledge through working together

1. Look at the diagram, figure 10.1, that shows a cross-section of the Earth.

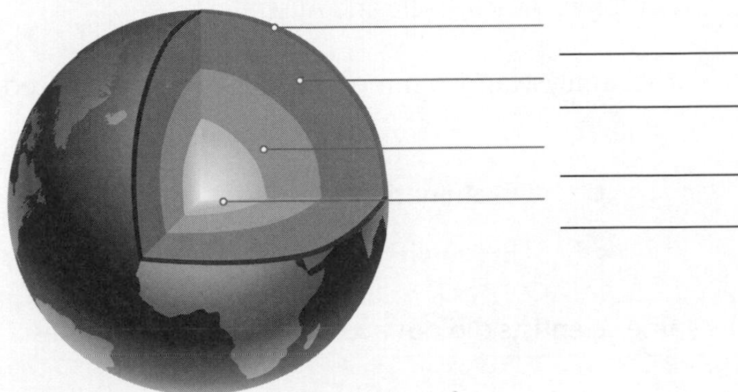

10.1

Complete the labels using the words from the list.

outer core	**inner core**	**crust**	**mantle**

2. The sentences describe how ideas about the Earth's crust have developed over time. They are not in the correct order of time. Write the numbers 1 to 5 in the boxes to show the correct order. The first one has been done for you.

[1] Map makers discovered that the shapes of two continents separated by an ocean appeared to match.

[ ] Scientists discovered that the position of Earth's magnetic north pole had changed over time.

[ ] A number of scientists developed the theory of plate tectonics.

[ ] Wegener suggested his hypothesis of continental drift.

[ ] von Humboldt suggested that the continents around the Atlantic Ocean had been joined together millions of years ago.

3. The sentences describe evidence that scientists have found about the structure of the Earth. Decide which evidence supports Wegener's idea of continental drift (W) and which evidence does not relate to continental drift (X). Write **W** or **X** in each box.

☐ **a** There are matching rock layers on the coasts of continents thousands of kilometres apart.

☐ **b** The average temperature of the Earth's atmosphere has been increasing over the last 200 years.

☐ **c** Some species of flying animals are found on more than one continent.

☐ **d** Identical fossils of animals that could not fly or swim long distances have been found in South America, Africa, India and Australia.

4. Name the process by which scientists analyse and comment on the papers produced by other scientists. Tick **one** box.

a Peer review ☐ **c** Examination ☐

b Assessment ☐ **d** Research ☐

5. What was the main reason why some scientists did not accept Wegener's hypothesis of continental drift? Tick **one** box.

a Wegener could not explain why identical fossils are found on different continents. ☐

b Wegener could not explain how rock layers matched on the coasts of different continents. ☐

c Wegener could not explain how the continents moved. ☐

d Wegener could not explain how the arrangement of different climate areas has changed over time. ☐

6. Sort the images showing the movement of tectonic plates into the correct order, from earliest to latest.

A B C

Order: ☐ → ☐ → ☐

10.2

7. Look at the diagram of the Earth's structure showing four tectonic plates.

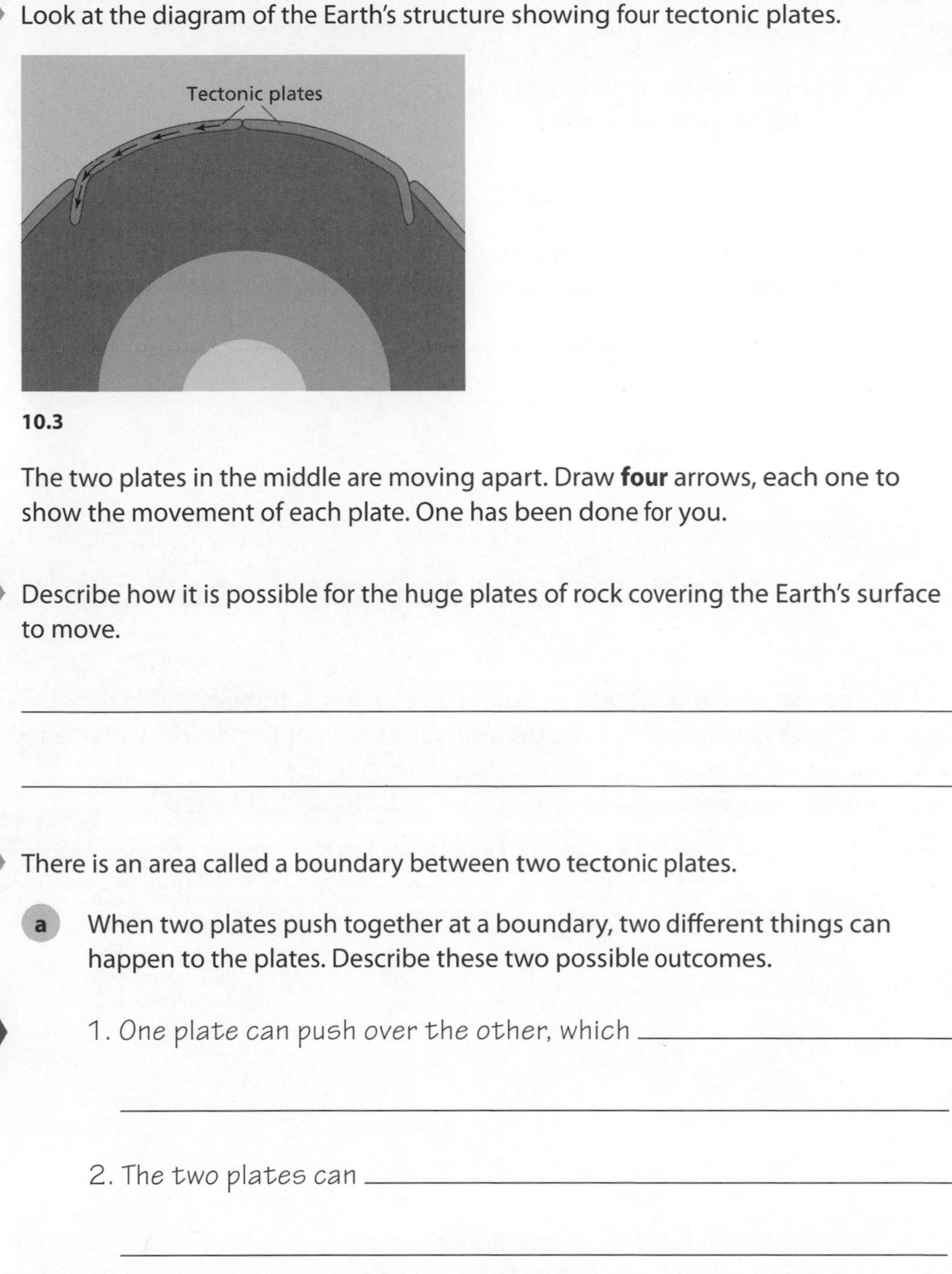

10.3

The two plates in the middle are moving apart. Draw **four** arrows, each one to show the movement of each plate. One has been done for you.

8. Describe how it is possible for the huge plates of rock covering the Earth's surface to move.

____

____

9. There is an area called a boundary between two tectonic plates.

a When two plates push together at a boundary, two different things can happen to the plates. Describe these two possible outcomes.

Show Me

1. One plate can push over the other, which ______________________

____.

2. The two plates can ______________________

____.

b Describe what takes place when two plates move apart from each other at a boundary.

____

____

10. **Challenge**

Plate tectonics is a model of what scientists think is happening to the Earth's surface. In order to test and improve models, scientists have to look for evidence.

a. Suggest **one** reason why plate tectonics is a better model than Wegener's original idea of continental drift.

____

b. A limitation of the plate tectonics model is that it is difficult to prove the process by which plates move over the mantle. Suggest **two** reasons why it is difficult.

____

____

10.2 Earthquakes, volcanoes and mountains

You will learn:

- To describe how the plate tectonics model explains earthquakes, volcanoes and fold mountains
- To describe how science can help predict earthquakes and volcanoes, and reduce their effects
- To understand some strengths and limitations of the plate tectonics model
- To describe how science is applied across societies and industries, and in research

1. Which of the following is NOT explained by the plate tectonics model?

a. Fold mountains

b. Ocean tides

c. Volcanoes

d. Earthquakes

2. Complete the sentences using the words from the box.

earthquake	**fold mountain**	**volcano**

When two tectonic plates collide and push together so that both plates rise up into the air, this makes a/an ____________________ .

When two tectonic plates that are pushed together suddenly slip past each other, this causes a/an ____________________ .

When two tectonic plates push together and molten rock rises through the crust, this causes a/an ________________________.

Questions **3** and **4** refer to the diagrams that show four different types of plate boundary.

A

B

C

D

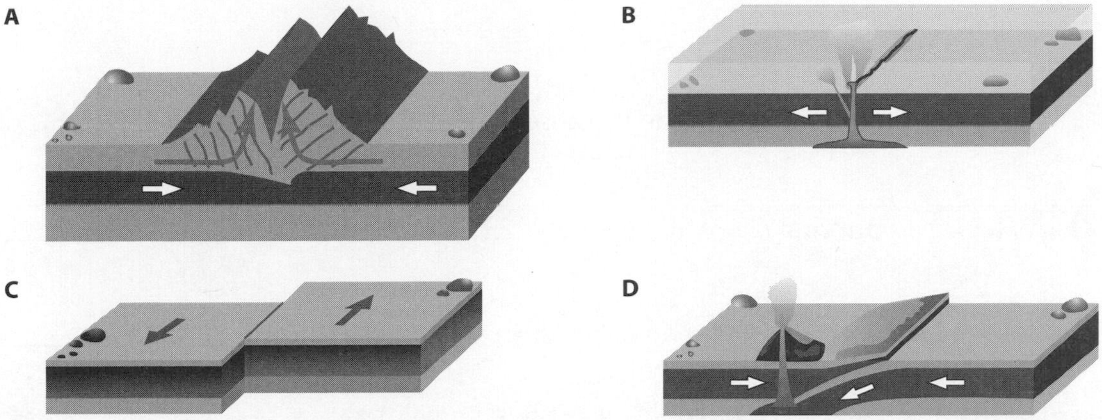

10.4

3. The diagrams show four different types of plate boundary. Describe what is happening to the tectonic plates in each case. The first one has been done for you.

A The plates are pushing towards each other, and the edges of both plates are being pushed upwards.

B ____

____

C ____

____

D ____

____

4. Name one example of an effect caused by each type of plate boundary – use the letters allocated in question 3. Use each word from the box **once**.

earthquakes	**volcanoes on land**	**volcanoes under ocean**	**fold mountains**

A ________________________

B ________________________

C ________________________

D ________________________

5. Describe how an earthquake can be caused at the boundary between two tectonic plates.

Show Me

At a plate boundary, two plates move ______________________________.

Over time, the force between the plates ______________________________.

Eventually, this force gets so large that the plates

______________________________.

6. The 'Ring of Fire' is a term used to describe an area near or at the edge of the Pacific Ocean, which runs along the coastlines of eastern Asia, western North America and western South America. There are many volcanoes along this 'Ring of Fire'.

a Suggest why there are so many volcanoes along this line.

____

____

b Suggest another type of natural event that will happen often along this line.

____

7. People who work in ocean-going ships, such as fishing boats and ocean liners, are not directly affected out at sea when the ground shakes during an earthquake. However, they should be prepared at all times and listen out for any news of earthquakes.

Explain why they should do this.

____

____

____

8. Practical

The diagram, figure 10.5, shows a very simple model of a volcano. You may have built one of these in class.

bottle partially filled with liquid
tray
'cone' of soil or sand

10.5

Suggest **three** limitations of this model.

____

____

____

9. Vulcanologists are scientists who study volcanoes. Describe **two** types of equipment they need to protect themselves when they are near the crater of an active volcano, and explain your answers.

____

____

____

____

10. Explain the differences between volcanoes and fold mountains.

____

____

____

____

11. **Challenge** Explain why it is so difficult for scientists to predict when the following events will occur.

a Volcanoes ____

____

b Tsunami ____

____

10.3 The Earth's atmosphere

You will learn:

- To know the percentages of the different gases in the Earth's atmosphere
- To know that the composition of the Earth's atmosphere can change because of natural processes and human activities
- To explain why accuracy and precision are important
- To present data using a pie chart
- To describe how climate modeling has led to worldwide changes in the use of particular substances
- To discuss how the uses of science can have a global environmental impact

1. Complete the sentences. Use the words in the box. You will need to use one term more than once.

oxygen	**water vapour**	**mixture**

Earth's atmosphere contains a ________________ of gases and ________________ .

If a sample of the atmosphere is described as 'dry air', this means it contains

no ______________________ .

We need one of the gases in the atmosphere to breathe. This gas is ______________ .

2. Match the gases to the percentages of those gases in dry air. Draw a line from each gas to the correct percentage.

nitrogen	just under 1%
argon	about 0.03%
oxygen	21%
carbon dioxide	78%

3. Practical

A scientist carried out an experiment to measure the percentage of nitrogen in the air. They repeated the experiment three times. Their results are shown in Table 10.1.

Repeat	1	2	3
Percentage of nitrogen (%)	73.5	74.5	74

Table 10.1

a Calculate the mean (average) percentage measured.

____

____

b These measurements are **precise**. Describe what precise means. Include the word 'range' in your answer.

Show Me

Results are precise if they are ________________.

This means all the values measured have a ________________.

c Are these results accurate? Justify your answer.

____

____

____

4. Describe **two** differences in the atmosphere of today compared with soon after the Earth formed, 4.5 billion years ago.

____

5. As life developed on Earth, the atmosphere changed. Name the process that takes place in plants that takes in carbon dioxide and gives out oxygen. Tick **one** box.

☐ **a** respiration

☐ **b** combustion

☐ **c** evaporation

☐ **d** photosynthesis

6. Complete the table. Name the three types of gases that are produced when fossil fuels are burnt, and match them to their effects on the atmosphere and living things.

Name of gas	Effects on atmosphere and living things
______________________	causes acid rain that damages plants and can cause breathing problems for some people
______________________	causes acid rain that damages plants and can cause breathing problems for some people
______________________	increasing amounts of this gas causes the Earth to warm up

Table 10.2

7. For each of the three gases given below, identify a natural process that can produce it.

Methane ______________________ Sulfur dioxide ______________________

Carbon dioxide ______________________

8. Look at the pie chart, figure 10.6.

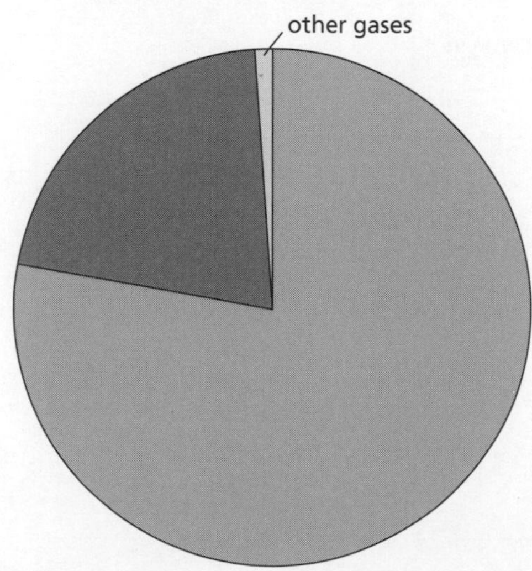

10.6

a Write the missing labels.

b Name **two** of the gases that are found in the segment labelled 'other gases'.

____

9. Complete the following article using words from the box. You do not need to use all the words.

CFCs	**oxygen**	**ozone**	**gases**	**liquids**	**banned**	**produced**

From the 1960s onwards, scientists have measured the levels of different __________ in the Earth's atmosphere. High up in the atmosphere, they discovered there is a layer of __________ gas, which protects the Earth from some damaging radiation from the Sun. The amount of this gas was reducing rapidly, due to humans' use of chemicals called __________. The protective layer is now being restored because all countries in the world __________ these harmful chemicals.

10. Practical

Describe an experiment that can be used to measure the amount of oxygen in a sample of air. Check that you have described:

- the equipment
- what substance or substances are needed (apart from air)
- the method used
- the chemical reaction that occurs
- the variables measured

____

____

____

____

____

____

____

____

11. Challenge

Scientists have determined that the average temperature of the Earth's atmosphere has been rising steadily over the past 150 years. They believe that most of this change is because of human activity. This is causing ice in glaciers and at the Earth's poles to melt more quickly than is natural.

a Name the gas that is mainly responsible for this change. ______________________

b Explain how the scientists have made the connection between the temperature rise in the past 150 years and human activity.

____

____

____

c Explain why the melting ice could speed up the rise in average temperature.

____

____

10.4 The water cycle

You will learn:

- To describe the processes in the water cycle
- To understand a model of the Earth's water cycle

1. Match the process to its definition. Draw a line from each process name to the correct definition.

evaporation	droplets of water in the air grow large enough so that they fall to the ground or ocean surface
condensation	water particles lose energy so they gather together to form droplets of water
precipitation	some water particles at the surface of a lake or ocean gain enough energy to break free from the surface

2. List **two** factors that can increase the rate at which water evaporates.

____

3. Which of the following are types of precipitation? Tick as many boxes as you think are correct.

☐ **a** snow
☐ **b** water vapour
☐ **c** rain
☐ **d** hail
☐ **e** steam

4. Explain what causes groundwater and water run-off to move from higher ground to lower ground.

____

____

5. **a** Describe groundwater and water run-off, including the differences between them.

____

____

____

____

b Why is it difficult to observe and measure groundwater?

____

6. **a** Sort the following statements about the water cycle into the correct order. Write a number from 1 to 4 in each of the boxes.

☐ **a** water vapour condenses to form small droplets of liquid water
☐ **b** water run-off forms streams and rivers that flow to the ocean
[1] **c** water evaporates from the surface of the ocean
☐ **d** water droplets fall as precipitation

b Groundwater forms at the same time as which one of the statements listed above? Write a, b, c or d ______________.

7. **a** Describe how a glacier forms.

____

____

____

b When a glacier melts, which part of the water cycle does it contribute to?

____

8. **Challenge** The diagram, figure 10.7, shows all the parts of the water cycle. Add arrows to the diagram to show how the cycle works. Make sure that each stage is connected into the cycle by at least one arrow.

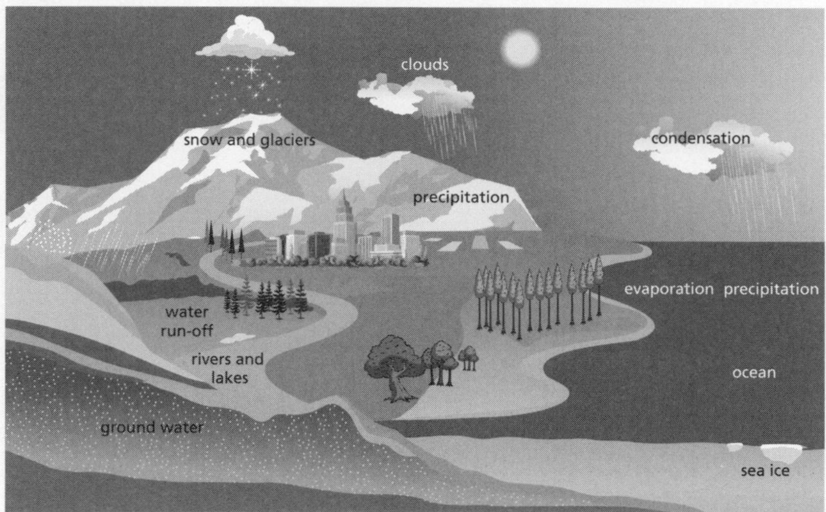

10.7 *The water cycle.*

Self-assessment

Tick the column which best describes what you know and what you are able to do.

What you should know:	I don't understand this yet	I need more practice	I understand this
The Earth's crust is divided into a number of tectonic plates that 'float' on the mantle			
Two tectonic plates may move apart from each other or push against each other			

Wegener suggested that the Earth's continents are moving apart			
It took a long time to find evidence to prove Wegener's idea			
Earthquakes and volcanic eruptions are sudden and violent events that can occur due to the forces between two plates at a plate boundary			
Two plates colliding over millions of years can form fold mountains			
The Earth's atmosphere is a mixture of 78% nitrogen, 21% oxygen and smaller amounts of gases including carbon dioxide and argon			
The mixture can change because of natural processes and human activities			
On Earth, water is evaporated from oceans, then condensed to form clouds and falls onto land or oceans as precipitation			
Precipitation that falls on land moves back to the oceans by groundwater or water run-off			
A combination of processes forms the water cycle, in which water moves from the oceans to the air, to the land and back to the oceans again			

You should be able to:	**I can't do this yet**	**I need more practice**	**I can do this by myself**
Describe how scientists discuss and improve their ideas over time			
Show how evidence can be used to prove a hypothesis			
Describe how different models can be used to show how plates pushing together can cause earthquakes, volcanoes and fold mountains			
Show how to model processes using diagrams and 3D models			
Suggest how to improve models			
Explain how accuracy and precision are different, and why they are important			

Suggest how to improve accuracy and precision in an investigation			
Describe investigations that are models for some parts of the water cycle			
Describe strengths and limitations of models of the water cycle			

If you have ticked 'I don't understand this yet' or 'I can't do this yet' or mostly 'I need more practice', have another look at the relevant pages in the Student's Book. Then make sure you have completed all the questions in this Workbook chapter and the review questions in the Student's Book. If you have already completed all the questions ask your teacher for help and suggestions on how to progress.

Teacher's comments

End-of-chapter questions

1. Name the scientific idea that offers the most complete explanation of how the Earth's crust changes over time. Tick **one** box.

- [ ] **a** continental drift
- [ ] **b** volcanoes
- [ ] **c** climate change
- [ ] **d** plate tectonics

2. San Francisco in California, USA, is on the West Coast of North America. Moscow, Russia is far inland, in the continent of Europe.

a Predict which city, San Francisco or Moscow, will experience a greater number and strength of earthquakes. ______

b Explain your answer to part **a**.

______

______

c The country of Nepal is inland, but can experience very strong earthquakes. Use your knowledge of plate tectonics to suggest why this is.

______

______

______

3. The island of Krakatoa, in Indonesia, is formed from the top part of a volcano. The lower part of the volcano is under the ocean. In August 1883, Krakatoa erupted. It is one of the largest eruptions for which we have records. Some of the explosions during the eruption were heard nearly 5 000 km away.

a Krakatoa is very close to the boundary between two tectonic plates. Suggest an explanation for the volcano being located there.

______

______

b The volcanic eruption was accompanied by other sudden movements of the Earth's surface. Name these movements.

______

It is estimated that more than 36 000 people were killed by the effects of the eruption.

c Some of these people were killed nearby, by the direct effects of the eruption. Suggest **three** different effects of a volcanic eruption that are dangerous.

______

______

______

d Many of the people were killed even though they lived on other islands, a long way from Krakatoa. Suggest an explanation for this.

4. Pollutant gases such as sulfur dioxide can be emitted into the atmosphere through human activities and natural processes.

a Suggest one natural source of sulfur dioxide gas.

b Name two other pollutant gases and state the human activity that produces them.

c Draw a simplified water cycle that includes the words 'condensation', 'evaporation', 'precipitation' and 'water runoff'. Add arrows to show how each of these words are linked in the cycle.

d Sulfur dioxide dissolves in rain water to create acid rain which can damage plants and trees. Suggest which stages of the water cycle can cause damage to the environment if the water contains acids.

- [ ] **a** precipitation
- [ ] **b** evaporation
- [ ] **c** water runoff
- [ ] **d** condensation

5. Challenge

Teams of scientists regularly measure the mixture of gases in the atmosphere in different parts of the world. Table 10.3 shows the results of measurements from three different areas.

Gas	% of total in area 1	% of total in area 2	% of total in area 3
X1	78.0	78.0	77.5
X2	21.0	20.8	20.5
argon	0.97	1.0	1.0
carbon dioxide	0.03 (estimate)	0.04 (estimate)	0.03 (estimate)
sulfur dioxide	0.00	0.016	0.00
Total	**100**	**100**	**99.03**

Table 10.3

a Name the gases X1 and X2.

X1 ________________, X2 ________________

b Suggest a reason why the levels of carbon dioxide are estimates.

____

c There is a pollutant gas in area 2. Name the gas and suggest where it could have come from.

____

d The total in area 3 does not add up to 100%. The scientists have checked the equipment and everything is working perfectly. Suggest what the 'missing' part of the mixture could contain, and explain your thinking.

____

____

One way of measuring the amounts of gases in a mixture is to measure the volume of a sample, then remove one of the gases from the mixture using a chemical reaction, and measure the volume of sample remaining.

e How can the results of this experiment be used to work out the volume of the reacted gas in the original sample?

____

____

f The amount of argon in a sample cannot be measured in this way. Suggest a reason why this is.

____

11.1 The planets and the Solar System

You will learn:

- To know that the force holding the components of the Solar System is gravity
- To describe how planets are formed

1. Complete the sentences using the words in the box.

Sun	**Earth**	**gravity**	**orbit**

Our Solar System contains eight planets that move around the ____________ .

Each planet is attracted to the other objects in the Solar System by the

force of ____________.

The nearly circular path that each planet follows is called its ____________ .

You will need this table of data about the planets in the Solar System to answer questions 2 to 4.

	Planet							
	Mercury	**Venus**	**Earth**	**Mars**	**Jupiter**	**Saturn**	**Uranus**	**Neptune**
Rotation period (hours)	1408	5833	23.9	24.6	9.9	10.7	17.2	16.1
Distance from Sun (millions of km)	58	108	150	228	779	1434	2873	4495
Orbit period (Earth days)	88.0	224.7	365.2	687	4331	10 747	30 589	59 800
Average surface temperature (°C)	167	464	15	–65	–110	–140	–195	–200
Number of moons	0	0	1	2	79	62	27	14

Table 11.1

2. Answer these questions about the planets.

a Name the planet with the largest number of moons. ______________________

b Name the planet that takes the longest time to spin once on its axis.

____

c Which planet is the hottest? ______________________

3. Look at the table.

a Describe the pattern in the distances of the planets from the Sun.

Show Me

From Mercury to Neptune (left to right in the table), the distance to the Sun ________________.

b Which other row in the table follows a similar pattern?

____

c Suggest why these two rows have a similar pattern of data.

Show Me

As the distance to the Sun ________________, it takes a planet longer to ________________ because ______________________________

4. Look at the table.

a Calculate how many times the Earth will orbit the Sun in the time it takes Saturn to complete one orbit. ______________________________

____

b Suggest why it is much colder on Saturn than it is on Earth.

____

____

5. Which planet was discovered by prediction? Tick **one** box.

- ☐ **a** Uranus
- ☐ **b** Jupiter
- ☐ **c** Mercury
- ☐ **d** Neptune

6. Which scientist developed the law of gravitation that tells us how planets orbit the Sun?

- [ ] **a** Newton
- [ ] **b** Le Verrier
- [ ] **c** Herschel
- [ ] **d** Galileo

7. Match the scientific terms to examples of them. Draw a line from each term to the correct example.

Hypothesis	Le Verrier calculated where Neptune could be found
Law	Observations and measurements of planets using telescopes
Prediction	Copernicus suggested that the Earth and other planets orbit the Sun
Evidence	Mathematical explanation of gravitation

8. **a** Name the cloud of dust and gas from which scientists think our Solar System formed. ______________________________

b Name the force that caused the dust and gas to gather into clumps.

____

c Describe how the Sun formed and started to produce light and heat.

____

____

____

d Explain why the Sun produces light but the planets do not.

____

____

9. Describe **three** major differences between the four inner planets of the Solar System (Mercury, Venus, Earth and Mars) and the four outer planets (Jupiter, Saturn, Uranus and Neptune).

____

____

____

10. Challenge Although we cannot go backwards in time to see how our Solar System formed, astronomers can still find evidence to support ideas about the process. Describe **two** different types of evidence that can be used in this way.

____

____

____

____

11.2 Tides

You will learn:

- To describe the causes of tidal forces on Earth
- To present measurements on a line graph
- To make a conclusion by interpreting results

1. What do tides mainly affect?

☐ **a** The orbit of the Moon

☐ **b** The orbit of the Earth

☐ **c** The water in Earth's oceans

☐ **d** The land on Earth's surface

2. Complete the sentences using the words from the box.

high tide	low tide	Earth	Moon	one	two	four

When the water level is down and we can see furthest down a beach, this is ______________ .

When the water level is up and we can see much less of a beach, this is ______________ .

The tide changes because of the position of the ______________ .

The number of high tides we have each day is ______________ .

3. The diagram, figure 11.1, shows the Earth and the Moon.

11.1

a Draw two arrows. One to show the force on the Earth caused by the Moon, and the other to show the force on the Moon caused by the Earth.

b Draw a shape around the Earth to show how the Moon affects the water on Earth.

4. Describe a 'tidal bulge'. ____

____

5. Describe how the Sun affects the tides on Earth. In your description, make sure you include a comparison of the size of the effect of the Sun with the effect of the Moon.

The Sun pulls on the water on Earth because ______________________ .

The size of the effect is much ______________________ .

When the Sun and Moon pull in the same direction, ______________________ .

When the Sun and Moon are at right angles, ______________________ .

6. How often do spring tides occur? Tick **one** box.

☐ **a** About twice a day

☐ **b** About twice a week

☐ **c** About twice every 28 days

☐ **d** Once a year, in spring

7. Decide whether each of the following sentences is true (**T**) or false (**F**). Write **T** or **F** in each box. The first sentence has been done for you.

a Spring tides occur when the Moon is Full or New. **T**

b Spring tides occur at the same time of the month as neap tides. ☐

c The difference in water height between low tide and high tide is smallest near the Earth's poles. ☐

d Tides are difficult to predict exactly. ☐

8. The Anand family have bought a small boat to use in the sea on their holidays. When they are not using it, they pull it up the beach so that it does not float away. Although it is small, it is heavy, so they don't want to have to pull it too far.

The diagram shows four different places they could leave the boat.

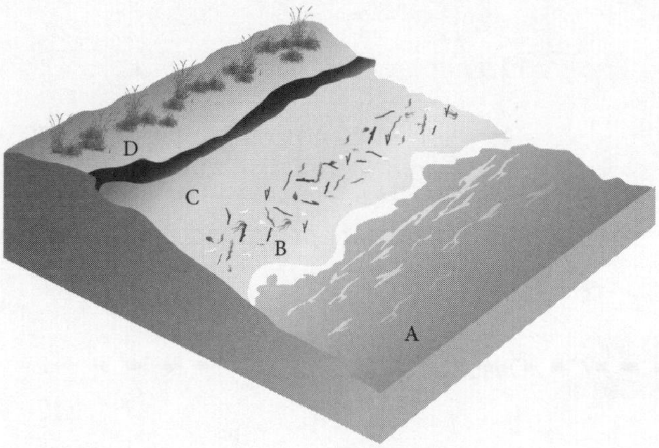

D
C
B
A

11.2

a Choose the best place for them to leave the boat, using your knowledge of tides. Tick **one** letter.

☐ **A**

☐ **B**

☐ **C**

☐ **D**

b Explain why each of the other places is not suitable.

____

____

____

9. **Challenge**

The graph, figure 11.3, shows how the water level at high tide changes over the course of one month, April.

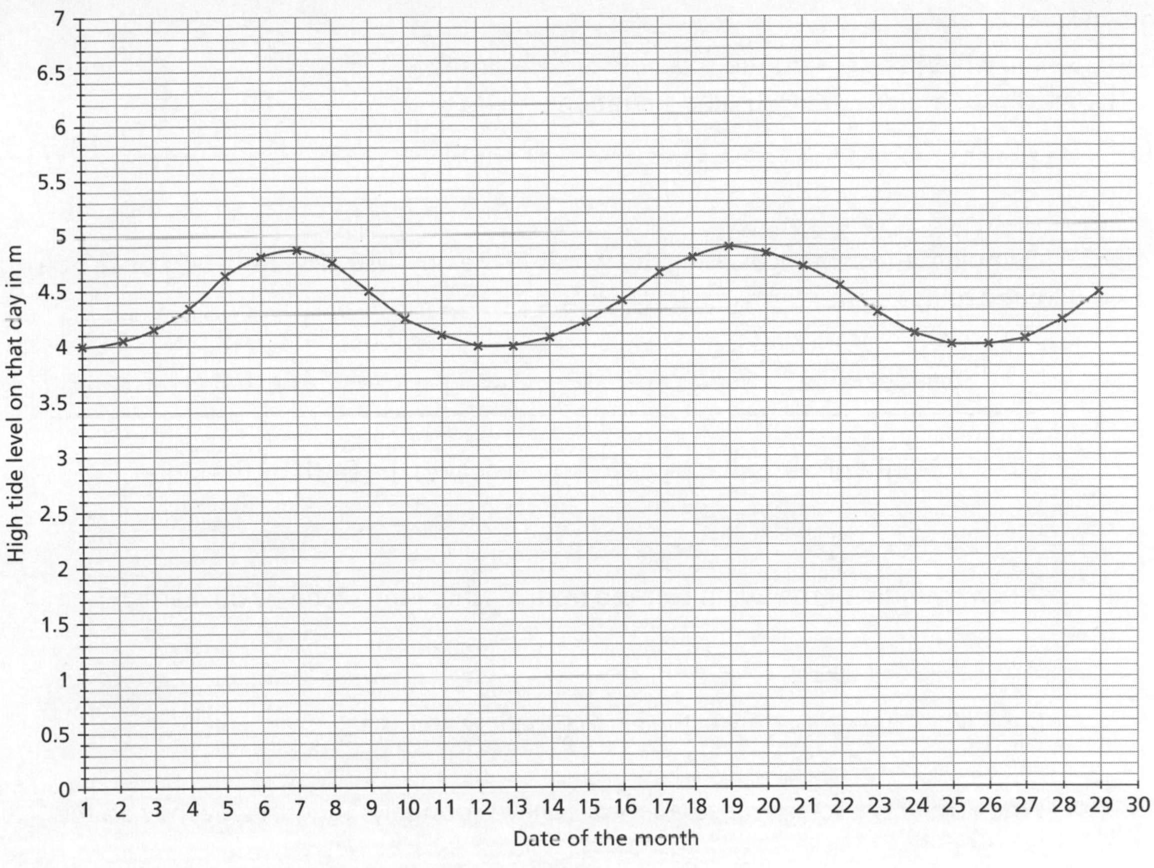

11.3

a Write down the date or dates at which a spring tide occurs.

____

b Calculate the difference in the height of high tide between a neap and a spring tide. Show your working.

____

____

c Predict the date of the next spring tide, after the graph ends. Show your thinking.

____

____

11.3 Eclipses

You will learn:

- To explain what happens during solar and lunar eclipses
- To describe a model's strengths and limitations

1. Complete the sentences using the words in the box. Use each term only once. You do not need to use all the terms.

lunar eclipse	**solar eclipse**	**total eclipse**	**partial eclipse**

When the light of the Sun is blocked from reaching the Earth by the Moon,

this is a ____________________.

When the light of the Sun is blocked from reaching the Moon by the Earth,

this is a ____________________.

If the Moon is completely in the Earth's shadow, this is a ____________________ .

2. When there is a solar eclipse, animals behave strangely. Suggest why they do this.

____

____

3. The diagram shows a type of eclipse.

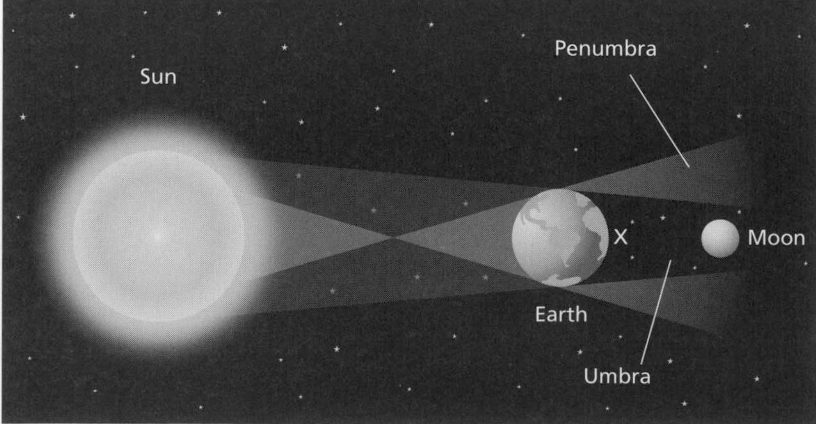

11.4

a Name the type of eclipse that is shown. ______________________

b Describe what a person would see if they stood at the point marked **X**.

____

____

4. The diagram shows a type of eclipse.

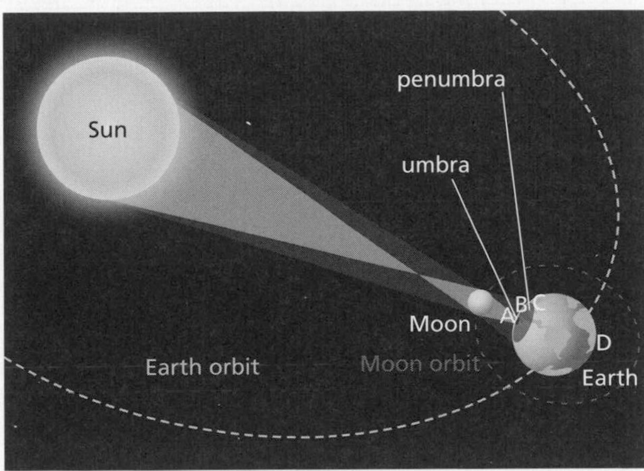

11.5

a Name the type of eclipse that is shown. ______________________

b At which point would a person have to be to see a total eclipse? Tick **one** box.

- [ ] **A**
- [ ] **C**
- [ ] **B**
- [ ] **D**

5. The pictures show two different views of the same solar eclipse, taken at the same time from different places.

11.6a *Total solar eclipse.*

11.6b *Partial solar eclipse.*

Explain why the people at these two different places see different things.

____

____

____

____

6. The Moon orbits the Earth once every 27.3 days. The Earth orbits the Sun once every 365.2 days (one year).

a Determine how many times the Moon orbits the Earth in one year. Show your working.

____

____

____

b There are only between two and five solar eclipses each year. Explain why the Moon does not eclipse the Sun every time it orbits the Earth.

____

____

7. Decide whether each of the following sentences is true (**T**) or false (**F**). Write **T** or **F** in each box. The first sentence has been done for you.

a A solar eclipse can only happen when there is a New Moon. [T]

b A lunar eclipse can only happen when there is a Quarter Moon. [ ]

c Every solar eclipse is a total eclipse when seen from the right place on the Earth. [ ]

d When a solar eclipse occurs, there must be a spring tide that day. [ ]

8. Yuri decides to make a model of a solar eclipse. He uses a torch, a table tennis ball and a small marble.

a Decide which of Yuri's items of equipment are best to model the Sun, the Earth and the Moon. Complete the table.

Equipment	What it models
torch	
small marble	
table tennis ball	

Table 11.2

b How should Yuri arrange the equipment to model the solar eclipse?

____

____

c Yuri says he wants to improve the model by making the distances to scale with real life. Explain why this is very difficult to do.

____

____

9. During a lunar eclipse, the Moon appears to be red or orange in colour. Explain why this happens.

____

____

10. Solar eclipses cannot be seen from the planet Venus. Explain why not.

Challenge

____

____

Self-assessment

Tick the column which best describes what you know and what you are able to do.

What you should know:	I don't understand this yet	I need more practice	I understand this
The Sun and planets formed from a solar nebula of gas and dust drawn together by the force of gravity			
The force of gravity holds planets in orbit around the Sun, and moons in orbit around planets			
Scientists predicted the position of Neptune and confirmed their prediction as telescope technology improved			
Tides are caused by the effects on ocean water of the force of gravity due to the Moon and the Earth's spin			
Spring tides occur when the force of gravity due to the Sun is in the same direction as that due to the Moon			
Neap tides occur when the force of gravity due to the Sun is at right angles to that due to the Moon			
A solar eclipse occurs when the Moon blocks the Sun's light from reaching the Earth's surface			
A lunar eclipse occurs when the Earth blocks the Sun's light from reaching the Moon's surface			
You should be able to:	**I can't do this yet**	**I need more practice**	**I can do this by myself**
Identify patterns in the data about planets			
Test and describe the accuracy of predictions using evidence			
Present data about tides in a line graph			
Write a conclusion based on the results of a graph			
Use a model to explain solar and lunar eclipses			
Describe a strength and a limitation of the eclipse model			

If you have ticked 'I don't understand this yet' or 'I can't do this yet' or mostly 'I need more practice', have another look at the relevant pages in the Student's Book. Then make sure you have completed all the questions in this Workbook chapter and the review questions in the Student's Book. If you have already completed all the questions ask your teacher for help and suggestions on how to progress.

Teacher's comments

End-of-chapter questions

1. Choose the correct order of planets, from smallest to largest.

☐ **a** Mercury, Earth, Uranus, Jupiter

☐ **b** Jupiter, Mercury, Uranus, Earth

☐ **c** Mercury, Uranus, Earth, Jupiter

☐ **d** Earth, Mercury, Jupiter, Uranus

2. Solar eclipses should only ever be viewed using special equipment.

a Explain why it is important never to look directly at the Sun, even during an eclipse.

____

____

b Describe some equipment that can be used to view the Sun safely during an eclipse.

____

____

____

3. There are eight planets in our Solar System that travel in orbits.

a What do all eight planets orbit around? ______________________

b Explain what an 'orbit' is.

____

____

c Name the force that causes the planets to travel in orbits. ____________

d Describe how scientists think the Solar System formed.

____

____

____

____

4. **a** The planets Mercury, Venus, Mars, Jupiter and Saturn were discovered by ancient civilisations that had no technology. Suggest why they could do this.

____

____

b Describe the technology that was developed, which allowed Herschel to discover Uranus.

____

____

c Tell the story of how Neptune was discovered.

____

____

____

5. The diagram, figure 11.7, shows a harbour wall and the measurement scale, which starts at 0 at the bottom of the wall and measures in metres. The graph, figure 11.8, shows the water level over the course of a day measured using this scale.

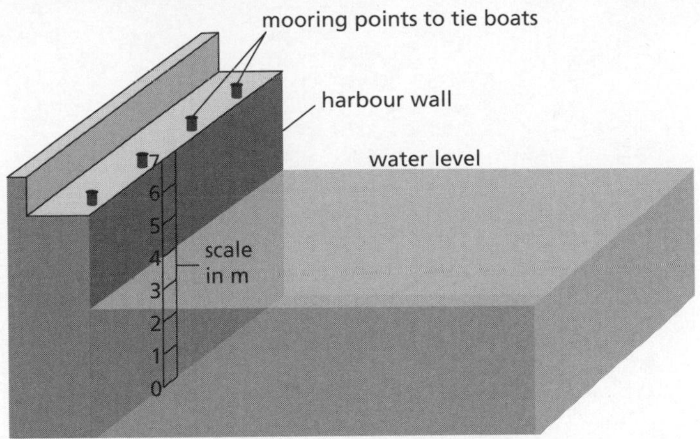

11.7 *Water level over 24 hours.*

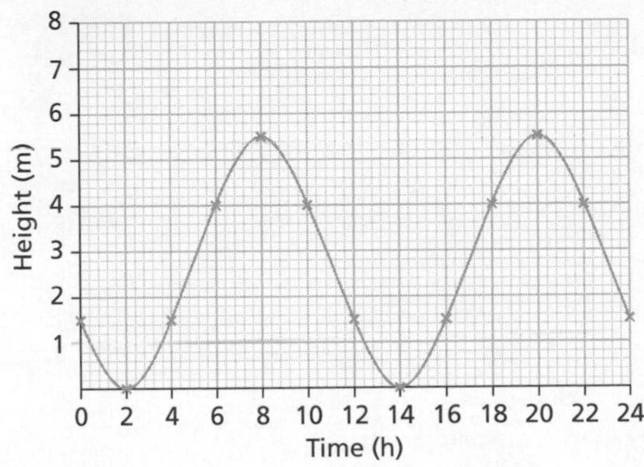

11.8

a Write down the height of the water level at low tide. ____________________

b Write down the height of the water level at high tide. ____________________

c Calculate the change in water level between low tide and high tide.

____

Challenge

A fisherman uses a rope to tie his boat to the top of the wall.
The rope is 6 metres long.

d Is his rope long enough? Explain your answer.

____

____

e What other information is needed to confirm your answer to part (d)?

____